Hybrid Communication Systems for Future 6G and Beyond

IEEE Press
445 Hoes Lane
Piscataway, NJ 08854

Hybrid Communication Systems for Future 6G and Beyond

Visible Light Communication & Radio over Fiber Technology

Rao Kashif
Faculty of Engineering & Computing
National University of Modern Languages, Pakistan

IEEE Press
WILEY

Published by John Wiley & Sons, Inc., Hoboken, New Jersey.
Published simultaneously in Canada.

Library of Congress Cataloging-in-Publication Data

Names: Kashif, Rao, author.
Title: Hybrid communication systems for future 6G and beyond : visible light communication & radio over fiber technology / Rao Kashif.
Description: Hoboken, New Jersey : Wiley, [2025] | Includes bibliographical references and index.
Identifiers: LCCN 2024035048 (print) | LCCN 2024035049 (ebook) | ISBN 9781394230280 (hardback) | ISBN 9781394230303 (adobe pdf) | ISBN 9781394230297 (epub)
Subjects: LCSH: Optical communications. | FiWi access networks. | 6G mobile communication systems.
Classification: LCC TK5103.59 .K367 2025 (print) | LCC TK5103.59 (ebook) | DDC 621.382/7–dc23/eng/20240819
LC record available at https://lccn.loc.gov/2024035048
LC ebook record available at https://lccn.loc.gov/2024035049

Cover Design: Wiley
Cover Image: © elen_studio/Shutterstock

Set in 9.5/12.5pt STIXTwoText by Straive, Chennai, India

To my beloved parents, siblings, and cherished wife,

Throughout every chapter of my life and in the creation of this work, your unwavering love, support, and encouragement have been my guiding lights. You have stood by me with boundless patience, understanding, and belief in my abilities, inspiring me to reach new heights and pursue my dreams relentlessly.

To my parents, your sacrifices and wisdom have paved the path for my journey, instilling in me values of perseverance, integrity, and compassion that continue to shape my character and endeavors.

To my siblings, your camaraderie and shared experiences have enriched my life immeasurably, reminding me of the importance of kinship and the strength found in unity.

And to my beloved wife, your unwavering love, patience, and unwavering belief in me have been my constant source of strength and inspiration. Your support has been my anchor, grounding me in times of uncertainty and propelling me forward with renewed determination.

This work is dedicated to you, with deepest love, gratitude, and appreciation for the profound impact you have had on my life and the creation of this endeavor.

Contents

About the Author

Dr. Rao Kashif is a distinguished figure in the field of wireless communication, renowned for his expertise in wireless optical system design, particularly in the domains of visible light communication (VLC) and radio-over-fiber (RoF) technology. He earned his PhD in wireless communication from the University of Science and Technology of China (USTC), a prestigious institution ranked among the top 100 universities worldwide by the QS World University Rankings.

With a wealth of experience and accolades to his name, Dr. Kashif has made significant contributions to the advancement of wireless communication technologies. During his tenure at Huawei Technologies, he served as a technical project manager, where his innovative approaches and leadership skills led to the successful implementation of numerous network designs. His exemplary performance earned him the Huawei Future Star Award, recognizing his outstanding contributions to the field.

Currently, Dr. Kashif holds the position of assistant professor and head of department at the National University of Modern Languages, Pakistan, within the Faculty of Engineering and Computing. In addition to his academic responsibilities, he served as the managing director of the Center to Advance Level Research and Development in Information and Communication Technology Research Center. In these roles, Dr. Kashif continues to drive cutting-edge research initiatives and mentor the next generation of engineers and researchers.

Dr. Kashif's research interests lie at the intersection of wireless communication and optical systems, with a particular focus on VLC and RoF technology. His groundbreaking work in these areas has garnered international recognition and has been published in esteemed journals and conferences.

Through his book, *Hybrid Communication Systems for Future 6G and Beyond*, Dr. Rao Kashif aims to share his expertise and insights with fellow researchers, engineers, and practitioners, providing a comprehensive resource for those seeking to explore the frontiers of indoor wireless networks.

Acknowledgments

First and foremost, I extend my heartfelt gratitude to Allah Almighty for granting me the strength and resilience throughout the process of writing this book. I am deeply indebted to my beloved parents and sibling for their unwavering support, encouragement, and understanding, which have been my pillars of strength.

I am immensely thankful to the National University of Modern Languages, Pakistan, for providing me with the conducive environment and resources essential for the fruition of this endeavor. The academic atmosphere fostered by NUML has been instrumental in shaping my thoughts and ideas.

I extend my sincere appreciation to the Center to Advance Level Research and Development (SMC-PVT) Ltd. for their invaluable experimental support, which significantly enriched the content of this work.

Special gratitude is extended to my alma mater, the University of Science and Technology of China, and the Micro-/Nano-Electronic Integration Center (MESIC) for their profound impact on my academic journey. I am particularly indebted to my esteemed PhD supervisor, Prof. Fujiang Lin, for his exceptional guidance, mentorship, and expertise in the realm of visible light communication. His unwavering support and encouragement have been instrumental in shaping my research endeavors and scholarly pursuits.

Introduction

In the ever-evolving landscape of wireless communication, the quest for efficient, reliable, and high-speed data transmission has become increasingly paramount. As the demand for seamless connectivity burgeons, particularly within indoor environments, novel technologies are essential to surmount the prevailing challenges and usher in a new era of communication infrastructures.

Hybrid Communication Systems for Future 6G and Beyond serves as a comprehensive exploration into the realm of cutting-edge communication technologies, with a particular focus on hybrid systems designed to optimize indoor wireless networks. This book delves into the convergence of various communication mediums, ranging from radio frequency (RF) and optical communication to visible light communication (VLC), to address the escalating data traffic demands in indoor environments.

The journey begins with an insightful overview of existing communication mediums, highlighting the potential of hybrid systems as a promising solution to overcome the limitations inherent in traditional wireless communication. This introductory section explores the exponential growth in indoor wireless communication and the necessity for novel technologies to manage the escalating data traffic. Emphasis is placed on the pivotal role of hybrid systems in optimizing indoor wireless communication networks, elucidating the synergy between various communication technologies to meet the evolving demands of modern communication infrastructures.

Subsequent sections delve into specialized domains, exploring advanced topics such as indoor VLC, radio-over-fiber systems, digital coherent integration, and proposed hybrid systems for indoor communication. These sections meticulously dissect the intricacies of system components, digital signal processing technologies, modulation schemes, and practical implementation considerations. Through comprehensive simulations, analysis, and real-world experiments, the book offers insights into the design and optimization of advanced communication infrastructures tailored for indoor environments.

With its interdisciplinary approach and forward-thinking perspectives, *Hybrid Communication Systems for Future 6G and Beyond* promises to be an invaluable resource for researchers, engineers, and practitioners striving to shape the future of wireless communication in indoor environments. By bridging the gap between theory and practice, this book provides a roadmap for the development and deployment of innovative communication solutions to meet the evolving demands of indoor wireless networks.

1

Introduction

1.1 Overview

Over the past 30 years, mobile networks have revolutionized the way in which we are connected. Mobile is and continues to be the world's most rapidly adopted technology, from education and entertainment to innovative new applications and services that are transforming how business is conducted around the world. Mobile is boosting our economies, transforming our communities, and empowering people in ways impossible to predict just a few years ago. The spectrum is the lifeblood that fuels this mobile connectivity; however, it is a finite resource. Every four years, the world's governments come together to agree on how the radio spectrum may be used during treaty negotiations of the World Radiocommunication Conference. It can take a decade before any new mobile spectrum can be used in new services and devices. With increasing smartphone adoption and advances in mobile technology enabling faster networks, people are using an ever-increasing amount of data. This always-on, always-connected demand will require access to a much wider mobile spectrum. The growth prediction of the Internet of Things (IoT) is illustrated in Figure 1.1. According to reference [1], there will be approximately 28 billion connected devices by 2021. The IoT will connect all indoor devices to the Internet, and we can imagine that the number of indoor communication devices will increase by a substantial amount.

1.2 Radio Frequency Communication

Radio frequency (RF) is a form of energy in the modification of time-dependent electronic and magnetic fields. In short, it is an electromagnetic wave that propagates readily in a vacuum, in space, or in solid-state media such as metal on a circuit board or through a coaxial cable, or it can propagate out of an antenna into space. In former times, people referred to it as a medium to carry

Hybrid Communication Systems for Future 6G and Beyond: Visible Light Communication & Radio over Fiber Technology, First Edition. Rao Kashif.

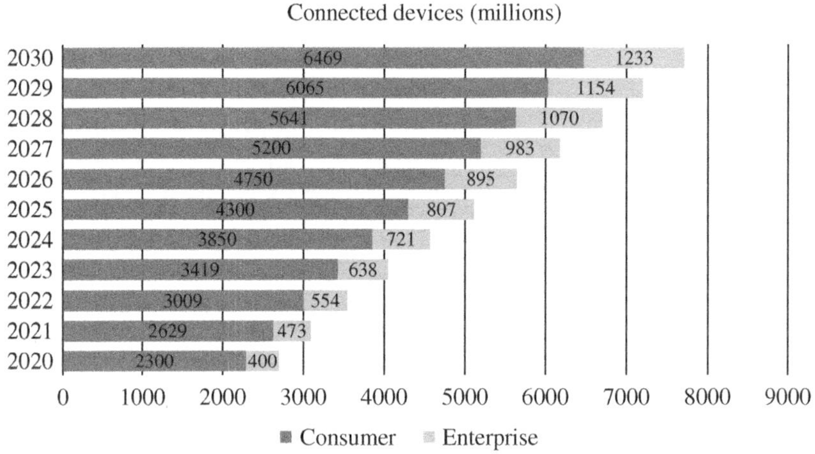

Figure 1.1 Growth of wireless connected devices.

electromagnetic waves; however, that notion was incorrect – it simply takes space. Our three-dimensional world carries these types of waves. We said RF is an electromagnetic wave; more specifically, only wave frequencies between 1 MHz and 3 GHz are called RF. Waves above 3 GHz and up to 30 GHz are referred to as microwaves. Higher frequencies between 30 and 300 GHz are termed as millimeter waves. Figure 1.2 [2] presents a chart of the electromagnetic spectrum. Along with frequency and wavelength, power is another crucial parameter to consider. The power of waves is directly proportional to the coverage. From cellular communication to Wi-Fi, Zigbee, BLE, and RF are serving the modern world in all fields of life. RF is currently the most widely used medium of communication. However, as we are moving toward faster and more secure communication for

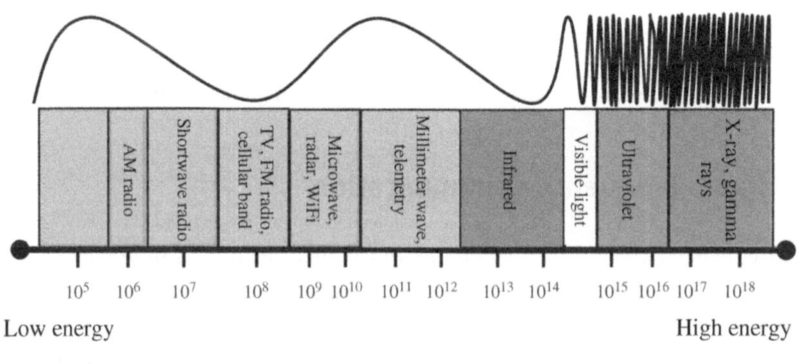

Figure 1.2 Electromagnetic spectrum chart.

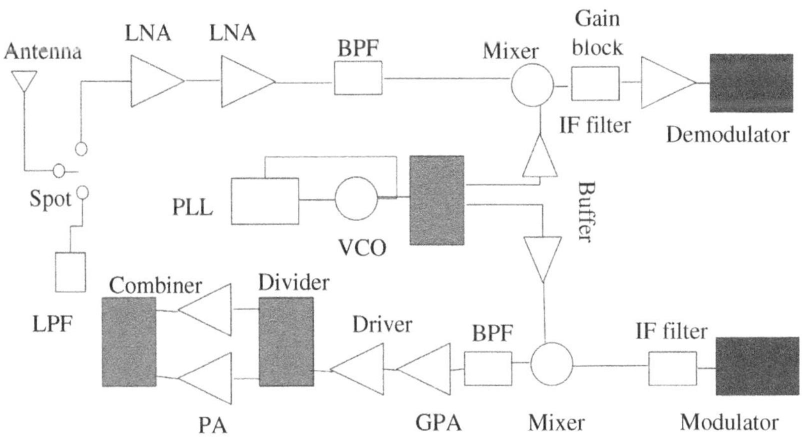

Figure 1.3 Radio frequency system block diagram.

the IoT and other applications, there are several issues that the RF industry is facing. Researchers are always trying to provide the best solutions with updates to the spectrum and technologies. Figure 1.3 illustrates a typical RF transceiver.

1.2.1 Limitations for Future RF Communication

In future 6G communication networks, RF communication may encounter several limitations, including.

1.2.1.1 Spectrum Congestion

With the increasing demand for wireless connectivity and the proliferation of connected devices, the available RF spectrum is becoming increasingly congested. This congestion can lead to interference and reduced network performance.

1.2.1.2 Limited Bandwidth

RF communication is constrained by the available bandwidth in the RF spectrum. As data rates continue to increase with emerging applications such as virtual reality, augmented reality, and high-definition video streaming, the limited bandwidth of RF communication may become a bottleneck.

1.2.1.3 Line-of-Sight Requirements

RF communication typically requires line-of-sight between the transmitter and receiver, particularly at higher frequencies. This can pose challenges in urban environments with obstacles such as buildings, foliage, and terrain, limiting the coverage and reliability of RF-based networks.

1.2.1.4 Signal Attenuation and Interference

RF signals are susceptible to attenuation and interference from environmental factors such as buildings, weather conditions, and electromagnetic noise. This can result in signal degradation and reduced network performance, particularly in outdoor environments.

1.2.1.5 Security Concerns

RF communication may be vulnerable to security threats such as eavesdropping, jamming, and unauthorized access. As cyberattacks become more sophisticated, ensuring the security and integrity of RF-based communication networks becomes increasingly challenging.

1.2.1.6 Energy Efficiency

RF communication technologies often require significant power consumption, particularly for long-range communication and high data rates. This can impact the energy efficiency of devices and infrastructure in future 6G networks, particularly in battery-powered devices and IoT deployments.

Overall, while RF communication has been the backbone of wireless networks for decades, addressing these limitations will be crucial for future 6G communication networks to meet the growing demand for high-speed, reliable, and secure wireless connectivity.

1.3 Optical Communication

Since the mid-1970s, optical fiber captured the interest of researchers. Due to limited bandwidth, sensitivity to crosstalk, radio and electrical interference, and security issues, researchers believed that copper would not be able to hold the future communication. The high bandwidth of a fiber channel offers more data flow compared to the other available communication mediums. As an increasing number of people utilize the Internet in their neighborhoods, the flow of data can slow to a trickle in comparison to the flow of data over fiber-optic broadband. Fiber optics technology is critical because only fiber optics can handle the increasing Internet bandwidth; technologies such as DSL and cable will not be able to keep up. Fiber is also a safe mode of transmission, and it is unaffected by environmental conditions. The cost of optical fiber continues to fall.

Similarly to other transmission systems, an optical fiber system also consists of a transmitter, a medium, and a receiver. Figure 1.4 presents a typical block diagram of the optical system. In the transmitter, a modulator converts the electrical signal to an optical signal. At the receiver, the light signal is downconverter to an electrical signal by a photodetector.

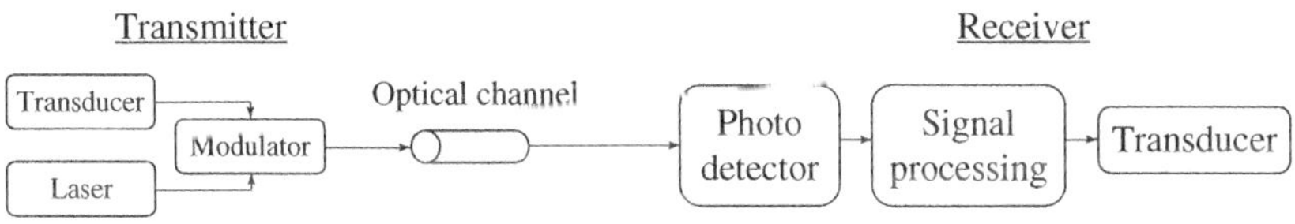

Figure 1.4 Optical system block diagram.

1.3.1 Future Opportunities for Optical Communication

In future 6G and beyond communication networks, optical communication technologies are poised to play a significant role and present several opportunities, including.

1.3.1.1 High Data Rates

Optical communication offers the potential for extremely high data rates, surpassing those achievable with traditional RF communication. This is especially crucial as future applications such as ultrahigh-definition video streaming, virtual reality, and augmented reality demand increasingly higher bandwidths.

1.3.1.2 Low Latency

Optical communication systems can provide lower latency compared to RF systems, which is essential for real-time applications like autonomous vehicles, telemedicine, and industrial automation. Reduced latency enhances responsiveness and improves overall user experience.

1.3.1.3 Large Bandwidth

Optical communication systems operate over a wide spectrum of optical frequencies, providing a vast bandwidth for data transmission. This large bandwidth enables the support of multiple users and devices simultaneously, facilitating the massive connectivity requirements of future networks.

1.3.1.4 Immunity to Electromagnetic Interference

Unlike RF communication, optical communication is immune to electromagnetic interference, making it ideal for deployment in environments with high electromagnetic noise levels, such as industrial facilities, urban areas, and dense urban environments.

1.3.1.5 Secure Communication

Optical communication offers inherent security advantages, as it is difficult to intercept or eavesdrop on optical signals without physical access to the transmission medium. This makes optical communication systems inherently more secure than RF systems, particularly for sensitive data transmissions.

1.3.1.6 Energy Efficiency

Optical communication systems can be more energy-efficient compared to RF systems, particularly for long-distance communication and high data rates. This is beneficial for reducing power consumption and extending the battery life of devices in future wireless networks.

Overall, optical communication technologies present significant opportunities for addressing the evolving requirements of future 6G and beyond communication networks, including high data rates, low latency, security, and energy efficiency. By leveraging these opportunities, optical communication can enable the development of more advanced and reliable wireless communication systems to meet the growing demands of next-generation applications and services.

1.4 Hybrid System

Two or more standards and systems are integrated into a hybrid system. Hybrid technology adds flexibility by using a combination of the most efficient technologies to meet the requirements of the modern world.

Each technology has advantages and disadvantages, and a hybrid system helps to overcome the cons of one technology with the pros of another technology. A prime example is combining wired and wireless technologies. A typical hybrid device provides network connectivity to both wireless and wired devices. Our wireless router is the most common example of a hybrid device that provides optical, Ethernet, and RF connectivity. Hybrid connectivity enables one to enjoy the privileges of all the systems that are integrated. For example, if someone requires fast file sharing, he can use an Ethernet connection with a router, and in the case of video streaming, he can switch to a wireless connection.

The integration of radio signals into optical technology offers numerous benefits and has been the focus of industry and researchers for three decades. Applications such as mobile network backhaul, indoor communication systems, and ultra-broadband wireless networks use microwave photonics hybrid technology. Low loss and broad bandwidth make the optical medium ideal for high traffic systems. Hybrid fiber radio (HFR) is an essential technology for wireless broadband and optical network integration, and it offers a wide range of services and applications. A simplified diagram of an optical and wireless hybrid network is presented in Figure 1.5.

1.4.1 Scope of Hybrid Communication

The scope of hybrid communication in the future is vast and encompasses a wide range of opportunities and applications. Hybrid communication refers to the integration of multiple communication technologies, such as optical, RF, and satellite

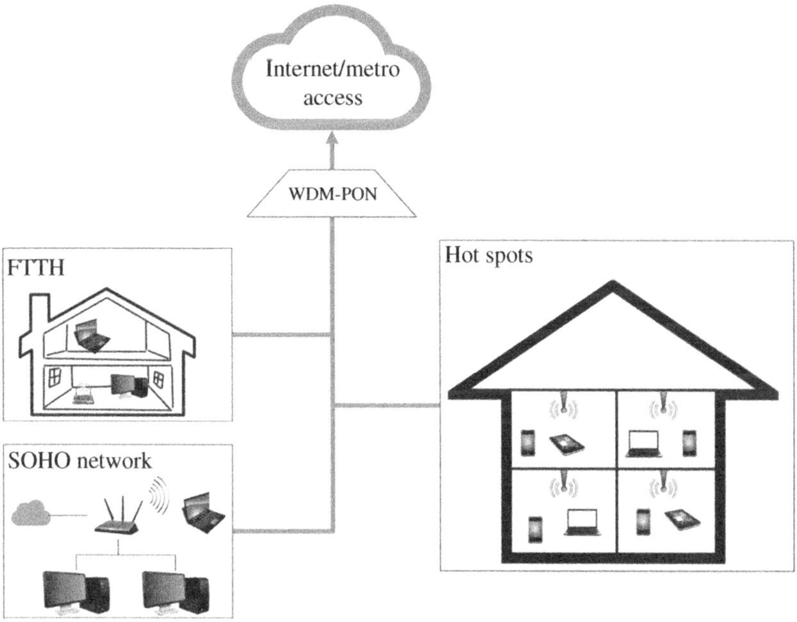

Figure 1.5 Optical and wireless hybrid network.

communication, to create robust, flexible, and high-performing communication networks. Some key aspects of the scope of hybrid communication in the future include.

1.4.1.1 Seamless Connectivity

Hybrid communication networks will enable seamless connectivity across diverse environments and scenarios, including urban, rural, and remote areas. By integrating multiple communication technologies, hybrid networks can provide continuous coverage and ensure connectivity in challenging environments.

1.4.1.2 Enhanced Reliability

Hybrid communication networks offer enhanced reliability and resilience by leveraging the strengths of different communication technologies. By combining optical, RF, and satellite communication, hybrid networks can mitigate the limitations and vulnerabilities of individual technologies, ensuring reliable and uninterrupted communication.

1.4.1.3 Improved Performance

Hybrid communication networks can deliver improved performance in terms of data rates, latency, and bandwidth compared to traditional single-technology networks. By intelligently routing traffic and dynamically allocating resources, hybrid networks can optimize performance and meet the diverse requirements of next-generation applications and services.

1.4.1.4 Flexibility and Scalability

Hybrid communication networks offer flexibility and scalability to accommodate evolving communication needs and requirements. With the ability to seamlessly integrate innovative technologies and adapt to changing conditions, hybrid networks can scale to support growing user demands and emerging applications.

1.4.1.5 Multimodal Communication

Hybrid communication networks enable multimodal communication, allowing devices and users to communicate using different technologies based on availability, capacity, and quality of service. This multimodal approach ensures optimal connectivity and enables seamless handover between different communication technologies.

1.4.1.6 Advanced Applications

The scope of hybrid communication in the future extends to enabling advanced applications and services across various sectors, including healthcare, transportation, agriculture, and smart cities. By providing reliable, high-speed connectivity, hybrid networks can support innovative applications such as telemedicine, autonomous vehicles, precision agriculture, and smart infrastructure.

1.5 History of Visible Light Communication

Visible light communication (VLC) has a rich history that spans centuries, from ancient signal fires and optical telegraphs to modern innovations in light emitting diode (LED) technology. The journey began with primitive methods of using light for communication and evolved significantly with breakthroughs such as Alexander Graham Bell's photophone. The modern era of VLC was propelled by advancements in LEDs and optical technologies, leading to the development of high-speed, secure communication systems that are now being integrated into next-generation networks like 6G.

1.5.1 Ancient Signaling Methods

Early civilizations utilized visible light for long-distance communication through methods such as signal fires, smoke signals, and torches. These primitive techniques laid the foundation for the concept of using light for communication purposes.

1.5.2 Optical Telegraphs

In the 18th and 19th centuries, optical telegraphs emerged as sophisticated systems for visual signaling over long distances. Claude Chappe's semaphore system, for instance, employed arrays of mechanical arms and flags arranged in specific configurations to convey messages rapidly across vast distances, introducing the concept of optical signaling.

1.5.3 Alexander Graham Bell's Photophone (1880)

Bell's invention of the photophone represented a significant advancement in optical communication. The photophone transmitted sound using modulated light beams, demonstrating the possibility of transmitting audio signals wirelessly through light. Although primarily an experimental device, the photophone laid the groundwork for future developments in optical communication technology.

1.5.4 Invention of Light Emitting Diodes (LEDs)

The invention and commercialization of practical LEDs in the 1960s and 1970s revolutionized lighting technology. LEDs offered numerous advantages over traditional incandescent and fluorescent bulbs, including higher efficiency, longer lifespan, and rapid switching capabilities. These characteristics made LEDs well-suited for applications beyond illumination, including data transmission.

1.5.5 Early Research into VLC (1990s–2000s)

In the late 20th and early 21st centuries, researchers began exploring the potential of using LED lighting for data communication. Early experiments demonstrated the feasibility of VLC for achieving high-speed data transfer rates. This period marked the beginning of systematic research efforts to develop VLC technology.

1.5.6 Harald Haas and Li-Fi (2011)

Professor Harald Haas's TED talk in 2011 introduced the term "Li-Fi" (light fidelity) and brought widespread attention to VLC technology. During his presentation, Haas demonstrated the transmission of data through modulated LED light, showcasing Li-Fi as a promising alternative to traditional RF-based wireless communication. Haas's talk ignited interest in VLC and accelerated research and development in the field.

1.5.7 Technological Advancements

Throughout the 2010s, significant advancements were made in VLC technology. These included improvements in LED efficiency, modulation techniques, and photodetector sensitivity, leading to higher data rates and more reliable communication. These technological advancements expanded the practical applications of VLC and contributed to its growing adoption.

1.5.8 Standardization Efforts

As VLC technology matured, efforts were made to establish standards for interoperability and compatibility. Organizations such as the Institute of Electrical and Electronics Engineers (IEEE) began developing standards for VLC communication protocols, ensuring consistency and facilitating broader adoption of the technology in various industries.

1.5.9 Integration with Modern Communication Systems

In recent years, VLC has been increasingly integrated into hybrid communication systems alongside traditional RF-based technologies. This integration allows VLC to complement existing communication infrastructure, offering benefits such as increased bandwidth, enhanced security, and reduced electromagnetic interference.

1.5.10 Current Trends and Future Prospects

VLC continues to be explored for a wide range of applications, including indoor positioning, smart lighting, vehicular communication, and underwater communication. Ongoing research aims to address remaining challenges, such as range limitations and interference management, to unlock VLC's full potential in next-generation communication systems. As VLC technology evolves, it is expected to play an increasingly significant role in shaping the future of wireless communication.

1.6 Visible Light Communication

Indoor VLC is in the initial stages of implementation. Chapter 2 provides an overview of numerous limiting factors in the practical implementation of VLC.

1.6.1 Problem 1

It is estimated that between 2015 and 2030, overall mobile communication around the world will octuple. Nowadays, for an indoor network, wireless fidelity (Wi-Fi) is the most popular wireless access technology. The most prominent standard is IEEE 802.11 n, supporting almost 600 Mb/s per access point. However, the attainable rate in this scenario is less conditional due to several factors. For

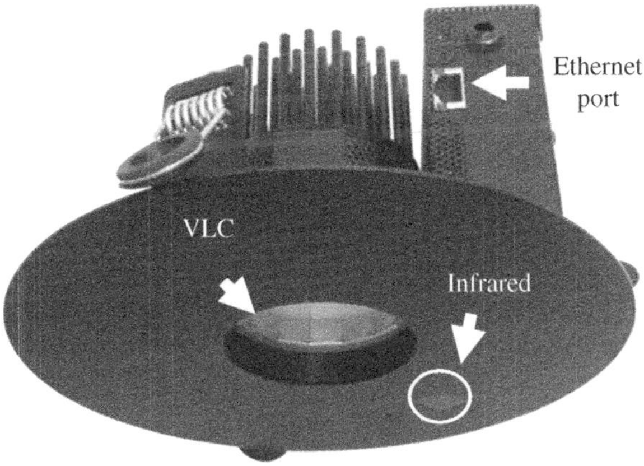

Figure 1.6 Pure Li-Fi device with Ethernet connection.

mobile access, long-term evolution (LTE) technologies are widely used in the world. Long-term evolution technologies, such as the 4th generation wireless technology, can provide 300 Mb/s at most if 4 × 4 multi-input and multi-output (MIMO) is employed. Due to weak capacity, the presently available RF standards are inadequate for 5G and further future networks. Limited bandwidth is one factor of low capacity in RF. As future indoor VLC systems or Li-Fi networks need a backhaul system that can hold high-speed communication. There is a need for a system that can connect an indoor high-speed system with a backhaul network, which can be a fiber to the home (FTTH) system.

1.6.1.1 Current Industry Trend

Few companies are currently providing visible light systems for indoor communication. All the available systems are based on an Ethernet connection. As the backhaul connection is already based on optical fiber, Ethernet-based systems cannot hold the high bandwidth communication that is required for future wireless communication. Figure 1.6 [3] illustrates a VLC module by Pure Li-Fi Company.

1.6.1.2 Possible Solution

The hybrid system provides high-speed optical connectivity to indoor VLC systems with ultra-high-speed optical backhaul networks. This high-speed connectivity will help the visible light system to work more efficiently.

1.6.2 Problem 2

A wireless communication network is incomplete without the facility of uplink communication. In light fidelity (Li-Fi), uplink requires that the transmitter and receiver maintain a directional link during transmission. It can significantly reduce the overall throughput of the network if both devices are constantly moving. So, in Li-Fi, the way in which the uplink traffic in a network will be operating is also a challenge. The RF and infrared (IR) can be considered for transmitting uplink data in Li-Fi networks; nevertheless, more innovative ideas are required to solve the uplink issues in Li-Fi networks. The three main hindrances to visible light uplinks are (i) the possibility of interference with the downlink signal; (ii) line of sight (LOS) issues between TX and RX; and (iii) power consumption in both scenarios if the uplink is through a device or a plugin device for uplink communication, as not all the devices currently support an IR module, which is the current solution in the market. A feasible solution is the hybridization of the visible light network with RF.

1.6.2.1 Current Industry Trend

Companies are providing IR as an uplink connection in a VLC system. However, IR itself is one of the hurdles in the implementation of the visible light system because our mobile devices and computers are not equipped with IR modules.

Only a few devices still have IR connectivity. To solve this issue, companies introduced a USB device, which an individual can plug into his or her communication device and gain IR uplink connectivity. Still, it is not convenient to plug in a device every time one wants to use the Internet.

Second, IR needs LOS connectivity; therefore, it is compulsory for the user to remain in one place to obtain uplink connectivity. Figure 1.7 displays one such device with an IR module in the market nowadays.

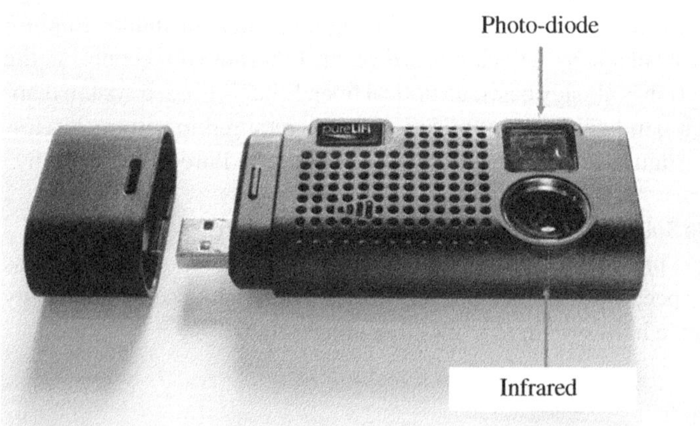

Figure 1.7 A USB device with IR connectivity. pureLiFi / https://factsc.com/li-fi/ / last accessed on May 02, 2024.

1.6.2.2 Possible Solution

The proposed hybrid system in this dissertation provides uplink connectivity through Wi-Fi. Also, the auto channel switching unit (ACSU) shifts the downloading traffic to Wi-Fi in the absence of a visible light channel. As Wi-Fi is easily accessible via all devices and there is also no issue with LOS in Wi-Fi, it will be a better solution than IR.

1.6.3 Opti Wave System Tool

It is a commercial optical design tool – an emerging leader in the development of optical links, components, systems, and fields of photonics nanotechnology, optoelectronic, optical networks, and other optical applications. The industrial clients of Opti System are Huawei, AT&T, and Lockheed Martin, among others.

Numerous systems designed in the Opti Wave system are published in IEEE journals and transactions such as MTT, Latin America Transactions, Photonics Journal, and Photonics Technology Letters.

References

1 Statista, "Number of IoT connected devices worldwide 2019–2023, with forecasts to 2030," 2023. [Online]. Available: https://www.statista.com/statistics/1183457

2 "Sura." 2008. [Online]. Available: www.sura.org

3 "Pure Li-Fi." 2013. [Online]. Available: www.purelifi.com

2

Visible Light Communication

2.1 Overview

We are consuming 100% more wireless data every year – this is exponential growth. In 2016, we consumed more wireless data than in the entire history of humanity. This raises the question of what will happen to security when the Internet of Things (IoT) becomes a reality, and everything becomes digital. There is an increasing demand for wireless data, and we need the energy to fuel all of them. In 2015, according to the center for energy-efficient communication, the world consumed 43 TWh just for accessing the wireless cloud – this was equal to the annual energy consumption of Rumania. However, there is another means of energy consumption, namely light. Light accounts for approximately 20% of all in-building electricity consumption; even with light emitting diodes (LEDs), that is a high amount of energy. What if data and light use the same power? Could we solve some future communication problems? Visible light can offer, among other things, unprecedented bandwidth, keep our data safe, and repurpose the energy that we use for illumination to provide wireless communication. This chapter highlights visible light communication (VLC) opportunities in future wireless communication, and it discusses the limiting factors that this medium of communication is facing in implementation.

VLC is all about using the light from LEDs as a medium of data transmission. LED lighting is used in many places, so people would be able to use a ubiquitous network. Imagine a world in which every LED light is an internet access point. Wi-Fi was an accidental invention in 1992 by Australian radio astronomer, Dr. John O'Sullivan, through a failed experiment to detect exploding mini black holes, and it has revolutionized digital communication. Wi-Fi currently makes up 60% of the global Internet traffic. Nevertheless, it has some issues; for example, it is sketchy at times, with a varying signal, and it is not secure.

Hybrid Communication Systems for Future 6G and Beyond: Visible Light Communication & Radio over Fiber Technology, First Edition. Rao Kashif.

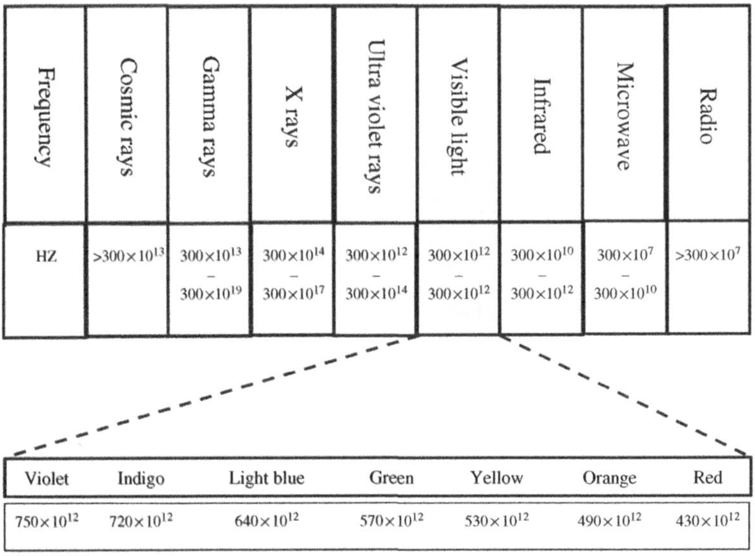

Figure 2.1 Visible light communication spectrum.

We must use something cheap, safe, plentiful, and more robust for data transfer. VLC can offer us fast and secure communication. Light is perfect because it is just part of the electromagnetic spectrum, as radio waves are, but at a much higher frequency – that is, the waves carry the energy cycle at a much faster rate than radio waves. Figure 2.1 presents the electromagnetic spectrum. Light also has a frequency range 10,000 times greater than radio waves, going up to 790 THz versus a maximum of 300 GHz for radio waves.

2.2 Background

For thousands of years back in history, the Native Americans and Romans used smoke or fire for message transfer [1]. In the 1880s, the Scottish scientist Alexander Graham Bell, with his assistant Charles Tainter, developed a photo phone, which can transmit sound signals on a beam of light [2]. This photo-phone is pictured in Figure 2.2 [3]. Thereafter, they began work on the microwave and then radio frequency (RF), which is the current wireless technology that we are using. Compared to microwaves and RF, the reported work in the field of visible light has been quite less in the past. LEDs for communication were considered as early as the 1990s, and infrared (IrDA) was formed in 1993. In 2001, researchers from Keio University started contributing to WLED, which led to broadband VLC [4]. The VLC Consortium (VLCC) was established in 2003 with the membership of

Figure 2.2 Bell's photo-phone. Bommervoice.ca / https://boomervoice.ca/explore-alexander-graham-bell-museum-beautiful-baddeck-cape-breton-island-nova-scotia-canada/ / last accessed on May 02, 2024.

famous Japanese companies such as Toshiba, Panasonic, Toyota, Mitsubishi, and NTT [5].

Around 2006, the Wireless World Research Forum's (WWRF) research related to VLC began to appear. Then, in 2007, NEC developed VLC, which Fuji television highlighted, and in the same year, Mostafa Afghani, who was the research assistant of Prof. Harald Haas, sent data using light signals. In 2008, Europe and the United States began to fund this research area after an invited paper published in Scientific American [6]. Home gigabit access (OMEGA), a European Union-funded project, also included optical wireless and VLC technology, and finally, in 2009, the IEEE called for 802.15.7 VLC. The integration of VLC into an indoor GPS was successfully demonstrated in Japan in 2010. In 2011, the free space optics device called Researchable Optical Near-joint Access (RONJA) was developed by the Czech Republic. In the same year, Dr. Harald HAAS presented Li-Fi his invention at a TED global talk [7]. Furthermore, much literature work is available in Refs. [2, 8–20].

2.3 VLC for Indoor Communication

The concept of light being used as a medium to transfer Internet data is referred to as Li-Fi. The term Li-Fi was first coined by the University of Edinburgh professor Harald Haas during a TED talk in 2011. He established a company in

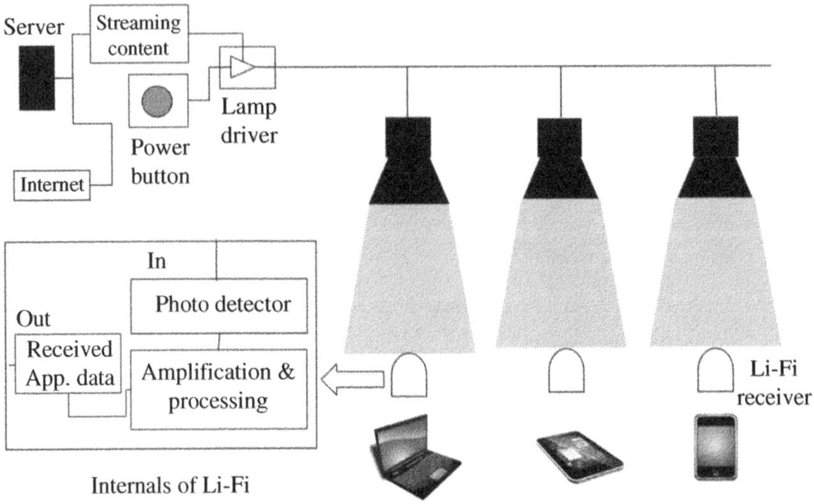

Figure 2.3 VLC architecture.

2012 called "Pure Li-Fi" with the aim to be the world leader in VLC technology. Researchers at Oxford University have published the results of visible light Internet at 223 Gb/s; with this speed, we can download 18 1.5-Gb movies in 1 second [18].

We can observe a cheap, light-powered Internet or Li-Fi everywhere, and this could easily lead to the IoT becoming a reality – a situation where all electronic devices communicate with each other. If Li-Fi becomes widely available, it would turn a corner into a technological revolution that would lead to a world in which the Internet and the transfer of information are no longer obstacles for the average user.

Li-Fi uses a photodetector to receive light signals and a signal processing element to convert the data into streamable content. An LED light bulb is a semiconductor light source, meaning that the constant current of electricity supplied to an LED light bulb can be modulated at high speeds without being visible to the human eye. The fluctuations represent 0 and 1 s.

Similarly to other wireless devices, a simple VLC or Li-Fi system consists of a transmitter and a receiver part. Figure 2.3 illustrates a typical VLC architecture. The transmitter has an LED driver, and the receiver is equipped with a photodiode and trans-impedance amplifier. Data are fed into an LED light bulb with signal processing technology. It then sends data, which are embedded in its beam, at high speeds to the photodetector.

The receiver then converts the tiny changes in the rapid dimming of LED bulbs into an electrical signal. Figure 2.4 portrays a Li-Fi bulb. The signal is then

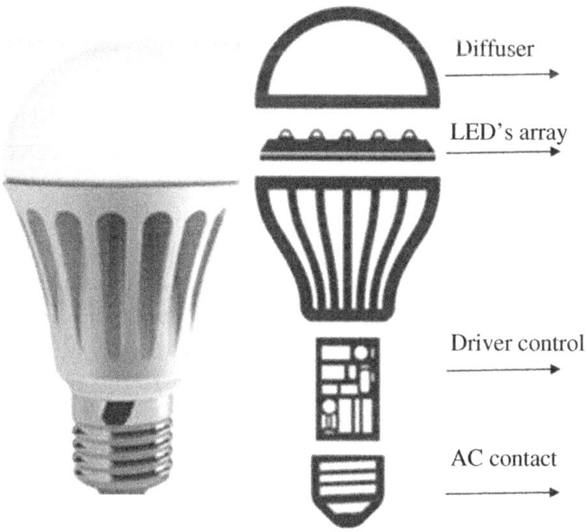

Figure 2.4 Li-Fi bulb.

converted back into a binary data stream, which we would recognize as web or audio applications that run on Internet-enabled devices.

2.4 Opportunities and Limitations

The question that arises here is why is visible light or Li-Fi technology necessary? If we look at the current RF technology, it has several problems. First, the future of RF is congested – we have a limited spectrum; for example, Wi-Fi uses 2.4 and 5 GHz bands, and this is going to be filled up since the number of wireless devices has exponentially increased in the past few years. Figure 2.5 [21] depicts the increase in indoor wireless communication in recent years. So, due to congestion, the technology does not work efficiently. The second problem is that we have limited bandwidth. Therefore, there are certain areas where this technology cannot be utilized; for example, in aviation and the health industry. Moreover, the most important problem is data security. There is a need for a secure method of communication, and the IoT is still in process because we need more secure systems for sending data. The following are some of the advantages that VLC has over RF.

i. High Speed: It is much faster than Wi-Fi. The tested speed is 10 Gb/s [22], and it can theoretically go up to 223 Gb/s [23]. VLC can achieve data transmission rates comparable to or even higher than those of traditional

RF communication systems. With advancements in LED technology and modulation techniques, VLC systems can support data rates in the gigabit-per-second range, enabling fast and efficient communication.

ii. Wide Spectrum: The visible light spectrum is vast and largely unregulated, providing ample bandwidth for VLC applications. This abundance of spectrum allows for the simultaneous transmission of multiple data streams without interference, enhancing the scalability and reliability of VLC systems.

iii. Utilization: It can be utilized in areas where RF cannot be implemented, especially inside aircrafts and in other RF-prohibited areas such as hospitals.

iv. Enhanced Security: VLC offers inherent security advantages due to its line-of-sight nature. Since light signals do not penetrate through walls or opaque obstacles, VLC communication remains confined to the physical boundaries of the transmission area, reducing the risk of unauthorized access or interception. Additionally, VLC systems can implement encryption techniques to further enhance data security.

v. Immunity to Electromagnetic Interference: Unlike RF communication, which is susceptible to interference from electronic devices and external sources, VLC operates in the optical domain and is immune to electromagnetic interference. This immunity ensures reliable communication in environments with high levels of electromagnetic noise, such as industrial settings and urban areas.

vi. Low Latency: VLC systems typically exhibit lower latency compared to RF communication, as light signals propagate at nearly the speed of light. This low latency is especially advantageous for applications requiring real-time communication, such as video streaming, online gaming, and industrial automation.

vii. Dual Functionality: VLC-enabled LED luminaires serve a dual purpose by providing both illumination and communication. This integration eliminates the need for separate infrastructure for lighting and communication, resulting in cost savings, simplified deployment, and enhanced energy efficiency.

viii. Environmental Friendliness: VLC technology leverages energy-efficient LEDs for both lighting and communication, contributing to environmental sustainability. By utilizing energy-efficient light sources, VLC systems reduce energy consumption and carbon emissions, making them environmentally friendly alternatives to traditional lighting and communication technologies.

ix. Compatibility with Existing Infrastructure: VLC can be seamlessly integrated into existing lighting infrastructure, leveraging the widespread deployment of LED lighting in various indoor and outdoor environments. This compatibility facilitates the deployment of VLC systems without the need for extensive infrastructure upgrades or modifications.

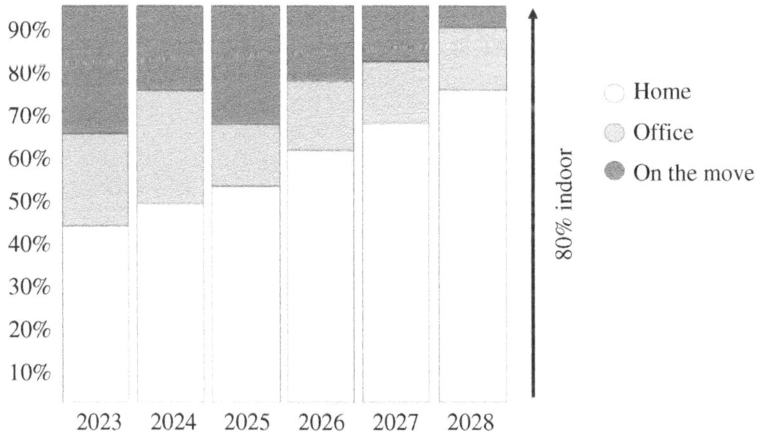

Figure 2.5 Increases in indoor wireless communication.

2.4.1 Applications

Low cost, high speed, and secure data transmission make VLC a favorite medium of communication. The following are few of the applications of VLC [24].

Light fidelity, an invention of Prof. Harald Haas, provides high-speed Internet connectivity for indoor communication. This technology can lead to the implementation of the IoT in the future [24, 25]. Li-Fi already achieved a speed of up to 10 Gb/s, which is much faster than Wi-Fi [26, 27].

i. Intra Vehicles: Visible light can be used for intra-vehicle and vehicle-to-roadside communication or road safety and traffic management. Park et al. [28] proposed a VLC-based intelligent traffic system.
ii. Hospitals: Visible light does not result in any electromagnetic interference and will not interfere with medical equipment [29]. HOSPI, a hospital transportation robot, was based on the visible light inside the building [30].
iii. Smart Signboards: Signboards are made of LEDs, and they can be used to transmit data, for example, at stations or airports. Ferreira et al. [31] proposed LED signboards for data transmission.
iv. Wireless Local Area Networks (WLANs): An LED-based WLAN can provide a high-speed Internet connection, as proposed in [32], that provides speeds of up to 10 Gb/s. In Ref. [33], a white LED-based VLC system was proposed in which IR was used for the uplink.
v. Limitations: To date, we have seen numerous opportunities and benefits of VLC. However, there are many limitations, which are hurdles in the implementation of this technology.

vi. Increasing Data Rate: LEDs were not made for data transmission in the past, so their bandwidth is less capable of meeting high-speed data transmission. Kodama and Haruyama [34] presented GaN-based micro-LEDs for VLC applications.

2.5 Modulation Techniques

Utilizing VLC offers a significant opportunity to enhance traditional RF communications, particularly in environments with short-range links such as homes or offices, where tasks like downlink Internet access, device-to-device file transfers, and video streaming are common. Just like conventional RF communications, various modulation schemes can be considered for VLC, each with its associated trade-offs in performance, cost, and complexity. However, unlike RF technologies, encoding VLC over amplitude and phase is not feasible in the traditional sense; instead, it relies on measuring the intensity of light waves, necessitating demodulation through the receiver's direct detection, also known as intensity modulation. With the RF spectrum facing increasing scarcity due to heavy congestion caused by surging demand for broadband communications and multimedia services, attention has shifted toward exploring alternative, highly efficient technologies. Exploiting the unregulated bandwidth available at optical frequencies makes optical wireless technologies particularly appealing for future local area networks. Furthermore, VLC systems offer a dual benefit by providing both illumination and wireless connectivity, addressing the growing need for indoor wireless communications and services.

In VLC systems, the optical output power of the source is adjusted based on the modulated signal. Various modulation schemes, including OOK (on–off keying), PWM (pulse width modulation), OFDM (orthogonal frequency division multiplexing), CSK (colour shift keying), GSSK (generalized space shift keying), and optical asymmetric modulation (OAM), have been proposed for use with VLC systems due to their enhanced performance and robust operation.

2.5.1 On–Off Keying

OOK modulation utilizes binary symbols, where '1' represents 'on' and '0' represents 'off'. However, when the 'off' mode is engaged, it does not necessarily mean that the LED is completely deactivated. This modulation scheme, known for its simplicity and ease of implementation, is commonly favored in VLC systems. One of the most straightforward methods for transmitting information involves generating light pulses and rapidly toggling the high-brightness LEDs on and off (referred to as pulse-based modulation schemes). White LEDs demonstrate faster response times compared to other LED types. It is worth noting that constructing an RGB white LED necessitates three independent driving circuits to control the white light.

vii. Pros
   i. Simplicity: OOK is one of the simplest modulation schemes, requiring straightforward operational mechanisms to function effectively.
   ii. Minimal Efficiency: OOK modulation is associated with minimal efficiency, making it easier to implement and operate.

viii. Cons
   i. Lower Data Rates: OOK provides lower data rates, particularly when maintaining different dimming values, which can limit its effectiveness in high-speed communication applications.
   ii. Decreased Efficiency with Dimming: The efficiency of OOK decreases as dimming occurs, potentially impacting its performance in varying lighting conditions.
   iii. Unchanging Modulation Intensity: The intensity of the ON/OFF modulations remains constant, which can hinder the communication process by limiting adaptability to changing environments or signal requirements.

2.5.2 Pulse Width Modulation

Within this modulation scheme, pulse width undergoes adjustments to accommodate different dimming levels while concurrently generating digital pulses crucial for modulation. Data transmission takes place during peak LED brightness, with the capability to fine-tune data rates in response to dimming requirements. Notably, researchers have demonstrated that the PWM frequency modulation technique facilitates dimming levels spanning from 0% to 100%, highlighting significant adaptability in regulating illumination levels.

ix. Pros
   i. Highest Dimming Level: The scheme achieves the highest dimming level without altering the intensity of light, eliminating the need to define on and off levels.

x. Cons
   i. Limited Data Rates: PWM is constrained by data rates limited to 4.8 kbps.

The authors in [34] proposed PWM-VLC using a DMD projector for voice guidance, demonstrating the transmission of four different fundamental waves simultaneously in various directions at high speed with a simple receiver design.

In [35], the impact of PWM dimming on a discrete multi-tone (DMT)-based VLC system's performance was analyzed through simulations, revealing that reliable communication requires PWM sampling of the DMT waveform at a rate at least twice the highest subcarrier frequency.

Pradana et al. [36] evaluated PWM accuracy, finding a 920-bps data transfer rate and 10^{-5}-bit error rate (BER) without affecting lighting function, based on parameters such as PWM frequency, transmitter-receiver distance, and receiver's angle of view.

2.5.3 Pulse Position Modulation (PPM)

In PPM modulation, the symbol time is divided into equal time slots, with one slot used to determine the pulse position. This technique suffers from low spectral efficiency and data rate, typically allowing for only one pulse per symbol duration. To address this limitation, a variation called overlap PPM (OPPM) has been proposed, enabling the transmission of multiple symbol durations within a single pulse.

Compared to traditional PPM, OPPM offers higher spectral efficiency and a wider range of dimming values, facilitating higher data rates. OPPM symbols have wider pulse widths, independent of dimming levels, requiring a fixed and small bandwidth for detection at the receiver across different dimming levels, unlike variable pulse modulation (VPM), which necessitates variable pulse widths for dimming.

Another variant, multiple-pulse PPM (MPPM), surpasses OPPM in spectral efficiency, making it more attractive than VOOK and variable pulse-position modulation (VPPM). As the code-word length increases, MPPM achieves higher spectral efficiency with less optical power.

The third type of PPM, VPPM, adjusts the duty cycles of optical symbols to encode bits. However, it is crucial to note that VPPM can change the duty cycle over time as the dimming level varies. Despite maintaining a constant bit rate regardless of dimming levels, the range diminishes as the light brightness decreases. The VPPM technique is illustrated below.

xi. Pros
   i. VPPM has the capacity to multiplex with varying levels of dimming within a given frame, aiding in achieving optimal resolution.
   ii. Offset-PPM demonstrated in [37] achieved a high speed of 11 Mbps over 1 m using a single white LED and a new code scheme, highlighting the potential for high-speed VLC systems.
   iii. PPSM (pulse position and pulse shape modulation) introduced in [38] combines pulse position and pulse shape modulation to increase transmission rates in VLC communications, enabling more efficient transmission at higher rates.
   iv. DiPPM (dicode pulse position modulation) implemented in [37] achieved a data rate of 13 Mbps in an indoor VLC system, demonstrating successful high-speed data transmission capabilities.

xii. Cons
   i. VPPM faces challenges in attaining optimal duty cycles, potentially impacting its efficiency and performance.
   ii. Single carrier modulation encounters high intersymbol interference (ISI) due to non-linear response to channel frequency in VLC systems, presenting constraints in data transmission efficiency.

2.5.4 Orthogonal Frequency Division Multiplexing

OFDM holds significant promise for VLC systems due to its ability to mitigate issues such as multipath propagation and ISI. By dividing the available spectrum into multiple orthogonal subcarriers, OFDM allows for efficient transmission of data over the optical channel. In VLC, where the bandwidth is limited, OFDM can help maximize spectral efficiency by enabling multiple subcarriers to transmit independent data streams simultaneously. Additionally, OFDM's resilience to frequency-selective fading makes it well-suited for indoor VLC environments where reflections and obstructions can cause signal degradation. With ongoing research and development efforts, OFDM continues to emerge as a key modulation scheme for enhancing the performance and reliability of VLC systems, paving the way for high-speed, robust wireless communication in indoor environments.

xiii. Pros
   i. Efficient Spectrum Utilization: OFDM divides the available spectrum into orthogonal subcarriers, allowing for efficient use of the bandwidth and maximizing data transmission rates.
   ii. Resilience to Interference: OFDM is robust against frequency-selective fading and multipath propagation, making it suitable for indoor VLC environments where signal degradation due to reflections and obstructions is common.
   iii. High Data Rates: By transmitting data simultaneously over multiple subcarriers, OFDM enables high-speed communication in VLC systems, supporting applications such as high-definition video streaming and real-time data transfer.
   iv. Flexibility: OFDM can adapt to varying channel conditions by adjusting the modulation and coding schemes on individual subcarriers, ensuring reliable communication in dynamic indoor environments.

xiv. Cons
   i. Complexity: Implementing OFDM in VLC systems requires sophisticated signal processing algorithms and hardware, which can increase system complexity and cost.
   ii. Sensitivity to Timing and Frequency Offset: OFDM requires precise synchronization and frequency alignment between the transmitter and

receiver, making it susceptible to timing and frequency synchronization errors.

iii. Power Efficiency: Transmitting over multiple subcarriers simultaneously may require higher power consumption, limiting the energy efficiency of OFDM-based VLC systems.

iv. Limited Range: While OFDM is effective for short-range communication in indoor environments, its performance may degrade over longer distances due to signal attenuation and dispersion.

2.5.5 Color Shift Keying

Color shift keying (CSK) is a modulation technique that leverages changes in the color of light to transmit data in VLC systems. In CSK, data is encoded by modulating the hue, saturation, or intensity of the transmitted light. Unlike traditional modulation techniques that rely on binary OOK, CSK utilizes a broader range of colors to represent multiple bits per symbol, thereby increasing the data transmission rate. This approach offers several advantages for VLC systems. Firstly, CSK allows for efficient spectrum utilization by leveraging the full range of colors available in visible light. Secondly, CSK is inherently robust against ambient light interference, as it operates by shifting the color of the transmitted light rather than toggling it on and off. Additionally, CSK can support multi-user communication by assigning distinct colors to different users or channels, enabling simultaneous data transmission without interference. Despite these benefits, CSK modulation requires careful calibration to ensure accurate color reproduction and decoding at the receiver. Furthermore, CSK performance may be affected by variations in ambient lighting conditions and the spectral response of the receiver. Overall, CSK holds promise as a modulation technique for VLC systems, offering a balance between spectral efficiency, robustness, and multiuser capability.

xv. Pros

i. Efficient Spectrum Utilization: CSK leverages the full range of colors available in visible light, allowing for efficient use of the spectrum and potentially higher data transmission rates.

ii. Robustness Against Ambient Light Interference: CSK modulation is less susceptible to ambient light interference compared to binary OOK techniques, as it operates by shifting the color of the transmitted light rather than toggling it on and off.

iii. Multiuser Communication: CSK can support multiuser communication by assigning distinct colors to different users or channels, enabling simultaneous data transmission without interference.

iv. Enhanced Security: CSK modulation can provide enhanced security in VLC systems by using a larger color space for data transmission, making it more difficult for unauthorized users to intercept or decode the transmitted data.

xvi. Cons
   i. Calibration Requirements: CSK modulation requires careful calibration to ensure accurate color reproduction and decoding at the receiver, which may add complexity to system implementation.
   ii. Sensitivity to Ambient Lighting Conditions: CSK performance may be affected by variations in ambient lighting conditions, potentially leading to fluctuations in signal quality and reliability.
   iii. Spectral Response Variation: CSK modulation may be impacted by variations in the spectral response of the receiver, which can affect the accuracy of data decoding and limit the achievable data rates.
   iv. Limited Range: Like other VLC modulation techniques, CSK may experience signal attenuation and dispersion over longer distances, limiting its effective range for communication.

2.5.6 Optical Asymmetric Modulation

OAM is a novel modulation technique specifically designed for VLC systems. Unlike traditional modulation techniques, which typically transmit data by modulating the intensity or color of light, OAM utilizes asymmetric waveforms to encode information. This asymmetry allows for the transmission of data through variations in the shape or duration of light pulses. OAM represents an innovative approach to data transmission in VLC systems, offering potential benefits in terms of high-speed communication and resilience to ambient light interference.

xvii. Pros
   i. High-Speed Data Transmission: OAM enables high-speed data transmission by utilizing asymmetric waveforms, potentially achieving faster data rates compared to traditional modulation techniques.
   ii. Resilience to Ambient Light Interference: OAM's reliance on waveform-based modulation rather than intensity or color modulation makes it less susceptible to fluctuations in ambient lighting conditions, ensuring reliable communication in diverse indoor environments.

xviii. Cons
   i. Complexity: Implementing OAM in VLC systems may require advanced signal processing techniques and specialized hardware, increasing system complexity and potentially raising costs.
   ii. Limited Research and Development: OAM is a new modulation technique in VLC, and more research and development are needed to fully understand its capabilities and limitations.

2.5.7 Discrete Multi-Tone (DMT)

DMT modulation, commonly known for its application in DSL (digital subscriber line) technology, has recently emerged as a promising modulation technique for VLC systems. DMT divides the available spectrum into multiple subchannels, each carrying its own data stream. This partitioning allows for adaptive modulation and coding on each subchannel, maximizing data rates and spectral efficiency. In VLC systems, DMT offers several advantages, including robustness against multipath propagation and interference, as well as flexibility in adapting to varying channel conditions. Additionally, DMT can support multi-user communication by allocating different subchannels to different users or channels, enabling simultaneous data transmission without interference. However, implementing DMT in VLC systems may require sophisticated signal processing algorithms and hardware, which can increase system complexity and cost. Despite these challenges, DMT holds promise as a modulation technique for VLC, offering high data rates, robustness, and flexibility for indoor wireless communication applications.

xix. Pros
   i. High Data Rates: DMT modulation allows for high-speed data transmission by dividing the available spectrum into multiple subchannels, each carrying its own data stream.
   ii. Robustness: DMT is robust against multipath propagation and interference, making it suitable for indoor VLC environments where signal degradation is common.
   iii. Adaptive Modulation and Coding: DMT supports adaptive modulation and coding on each subchannel, optimizing data rates and spectral efficiency based on channel conditions.
   iv. Multiuser Communication: DMT can support multi-user communication by allocating different subchannels to different users or channels, enabling simultaneous data transmission without interference.

xx. Cons
   i. Complexity: Implementing DMT in VLC systems may require sophisticated signal processing algorithms and hardware, increasing system complexity and potentially raising costs.
   ii. Sensitivity to Channel Variations: DMT performance may be affected by variations in channel conditions, requiring adaptive modulation and coding schemes to maintain optimal data rates.
   iii. Limited Range: Like other VLC modulation techniques, DMT may experience signal attenuation and dispersion over longer distances, limiting its effective range for communication.

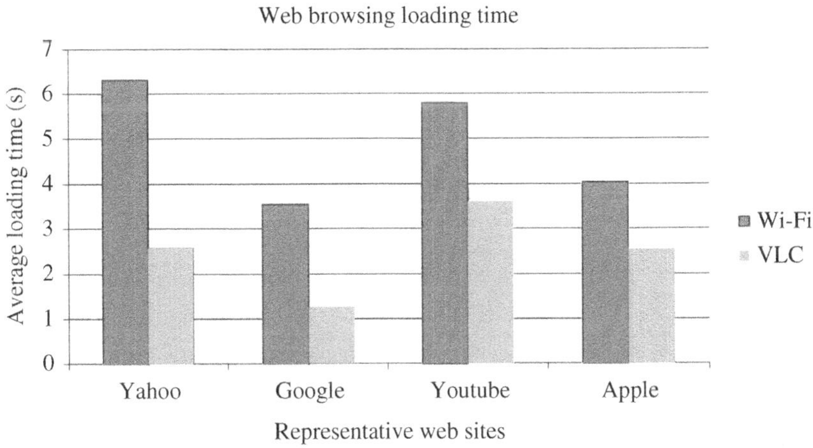

Figure 2.6 Li-Fi & Wi-Fi comparison.

2.6 Light Fidelity and Wireless Fidelity Comparison

Light fidelity is a light-based communication technology that delivers high-speed, bidirectional network, mobile communication in a manner like Wi-Fi. Instead of a Wi-Fi modem, Li-Fi uses transceiver-fitted LEDs that can light a room as well as transmit and receive information. In general, Wi-Fi offers wireless coverage within a building, whereas Li-Fi offers high-density wireless data coverage in a confined area and relieves radio interference issues. Li-Fi has higher speeds than Wi-Fi – 10,000 times the frequency spectrum of radio. Li-Fi technology is used in airlines, underwater explorations, and operation theatres in hospitals, offices, and home premises for data transfer and Internet browsing, whereas Wi-Fi is used for Internet browsing with the help of Wi-Fi hotspots. Ntogari et al. [35] and Pradana et al. [36] provides experimental comparisons of Wi-Fi and Li-Fi systems, and a comparison of the throughput achieved and the loading time in browsing is presented in Figures 2.6 and 2.7.

2.7 VLC Transmitter and Receiver

A VLC system utilizes visible light, typically emitted by LEDs, to transmit data wirelessly. In a VLC system, data is modulated onto the light signal, and receivers equipped with light sensors capture the modulated light and decode the data. Here is a brief overview of how a VLC system works.

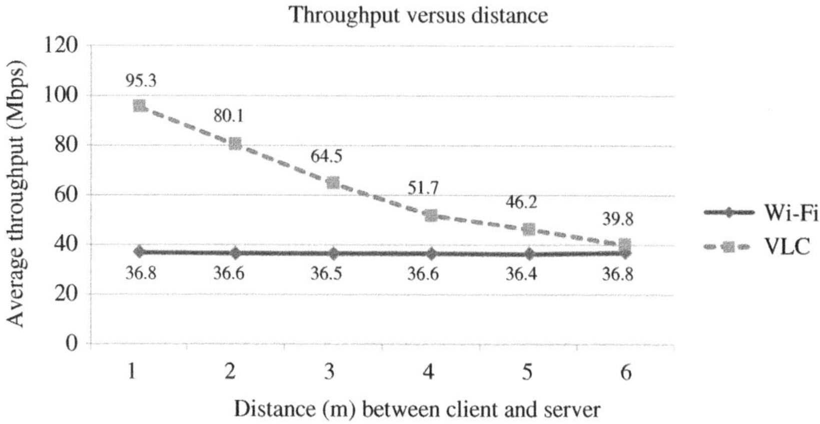

Figure 2.7 Li-Fi & Wi-Fi throughput versus distance.

i. Transmitter: The transmitter consists of a light source, such as an LED, and modulation circuitry to encode data into the light signal. Various modulation techniques like OOK, PWM, and OFDM can be used to modulate the light signal.
ii. Transmission Medium: The light signal travels through the air, transmitting data over short to medium distances. The transmission medium can be indoor spaces like offices, homes, or outdoor environments with line-of-sight communication.
iii. Receiver: The receiver includes a light sensor, such as a photodiode or photodetector, to capture the modulated light signal. The received light signal is then demodulated to recover the original data using similar modulation techniques employed at the transmitter.
iv. Data Processing: Once the data is demodulated at the receiver, it undergoes further processing to extract the transmitted information. This may involve error correction coding, data framing, and other signal processing techniques to ensure reliable communication.

2.7.1 VLC Transmitter

A VLC transmitter is a device designed to convert electrical signals carrying data into optical signals, typically using LEDs or laser diodes, and transmit them through the visible light spectrum. The transmitter consists of several key components.

i. Light Source: The primary component of a VLC transmitter is the light source, which emits light in the visible spectrum. This can be an LED or a laser diode, depending on the application and requirements of the system.

ii. Modulation Circuitry: The modulation circuitry encodes data onto the light signal. Various modulation techniques can be used, including OOK, PWM, OFDM, and CSK, among others.
iii. Driver Circuit: The driver circuit provides the necessary electrical signals to control the intensity, color, and modulation of the light source. It may include amplifiers, filters, and other components to ensure the desired characteristics of the transmitted light signal.
iv. Optical Components: Optical components such as lenses, collimators, and diffusers may be used to shape and direct the light signal emitted by the light source. These components help optimize the transmission efficiency and coverage area of the VLC system.
v. Control Interface: The control interface allows for communication with external devices and systems to coordinate the operation of the VLC transmitter. It may include interfaces such as USB, Ethernet, or wireless communication protocols for remote control and monitoring.
vi. Power Supply: A power supply unit provides the necessary electrical power to operate the VLC transmitter components. Depending on the application, the power supply may be mains-powered, battery-powered, or powered by renewable energy sources such as solar panels.

2.7.2 VLC Receiver

A VLC receiver is a device designed to capture and decode modulated light signals transmitted by a VLC transmitter. Below are the components and functionality of a VLC receiver.

i. Light Sensor: The primary component of a VLC receiver is the light sensor, typically a photodiode or photodetector. The light sensor detects the modulated light signal emitted by the VLC transmitter and converts it into an electrical signal for further processing.
ii. Amplification Circuitry: The electrical signal generated by the light sensor is often weak and requires amplification to enhance its strength for accurate signal detection. Amplification circuitry is used to boost the signal to a usable level.
iii. Demodulation Circuitry: The modulated light signal contains encoded data, which needs to be demodulated to recover the original transmitted information. Demodulation circuitry processes the amplified electrical signal to extract the data, using techniques such as OOK, PWM, or OFDM, depending on the modulation scheme employed by the VLC transmitter.
iv. Signal Processing: Once the data is demodulated, it undergoes further signal processing to extract the transmitted information accurately. This may involve

error correction coding, data framing, and other processing techniques to ensure reliable communication.

v. Data Output: The processed data is then outputted from the VLC receiver for use by external devices or systems. This may involve transmitting the data to a computer, smartphone, or other communication devices for further processing or display.

References

1 Haruyama, S. (2010, September). Visible light communications. In *36th European Conference and Exhibition on Optical Communication* (pp. 1–22). IEEE.

2 B. M. Conlon, "Patent application publication (10) Pub. No.: US 2008/0125265 A1 Eg: Patent application publication," vol. 1, no. 19, 2008.

3 "Bells Lab Technical Museum." Available: https://www.bell-labs.com.

4 Y. Tanakat et al., "Indoor visible communication utilizing plural white LEDs as lighting."

5 M. Kavehrad, "Sustainable energy-efficient wireless applications using light," *Consum. Commun. Networking, IEEE*, vol. 48, no. December, pp. 66–73, 2010.

6 M. Kavehrad, "Broadband room service by light," *Sci. Am.*, vol. 297, no. 1, pp. 82–87, 2007.

7 H. Haas, "Li-Fi consortium," *TED*, 2011. [Online]. Available: http://www.lificonsortium.org/.

8 T. Komine and M. Nakagawa, "Fundamental analysis for visible-light communication system using LED lights," *IEEE Trans. Consum. Electron.*, vol. 50, no. 1, pp. 100–107, 2004.

9 S. Haruyama, "Visible light communication using sustainable LED lights," *2013 Proceedings of ITU Kaleidoscope: Building Sustainable Communities*, pp. 1–6, 2013.

10 D. Shin et al., "Visible light communication: Tutorial," *Tutorial*, no. March, pp. 1–78, 2008.

11 S. Haruyama, "Progress of visible light communication," *Communication*, pp. 3–5, 2010.

12 S. Haruyama, "Progress of visible light communication," in *Optical Fiber Communication Conference, OSA Technical Digest (CD)* (Optica Publishing Group, 2010), paper OThH2.

13 M. Z. Afgani, H. Haas, H. Elgala, and D. Knipp, "Visible light communication using OFDM," *2nd International Conference on Testbeds and Research Infrastructures for the Development of Networks and Communities, 2006. TRIDENTCOM 2006*, pp. 129–134, 2006.

14 C. G. Lee, "Visible light communication," in *Visible Light Communication, Advanced Trends in Wireless Communications*, Wiley-WWRF Series Dr., M. Khatib (Ed.), pp. 32–35, 2011.

15 P. Salian, S. Prabhu, P. Amin, S. Naik, and M. Parashuram, "Visible light communication," *2013 Texas Instruments India Educators' Conference*, pp. 379–383, 2013.

16 M. Nakajima and S. Haruyama, "New indoor navigation system for visually impaired people using visible light communication," *Eurasip J. Wirel. Commun. Netw.*, vol. 2013, no. 1, 2013.

17 S. Arnon, "Visible light communication," *Can. Med. Assoc. J.*, vol. 130, no. August, p. 557, 1984.

18 A. Cuthbertson, "LiFi internet breakthrough: 224Gbps connection broadcast with an LED bulb," 2015. [Online]. Available: http://www.ibtimes.co.uk/lifi-internet-breakthrough-224gbps-connection-broadcast-led-bulb-1488204.

19 "Indoor wireless communication," *Informa.* [Online]. Available: https://informa.com/.

20 A. Cuthbertson, "LiFi internet: First real-world usage boasts speed 100 times faster than WiFi," 2015. [Online]. Available: http://www.ibtimes.co.uk/lifi-internet-first-real-world-usage-boasts-speed-100-times-faster-wifi-1530021.

21 Latif Ullah Khan, "Visible light communication: Applications, architecture, standardization and research challenges," *Digit. Commun. Netw.*, vol. 3, pp. 78–88, 2016.

22 H. Haas and C. Chen, "What is Li-Fi?," *J. Light. Technol.*, no. 99, pp. 3–5, 2015.

23 "Li-Fi to speed up IoT? | the internet of things," 2013. [Online]. Available: www.theinternetofthings.eu/li-fi-speed-iot.

24 J. Vincent, "Li-Fi revolution: internet connections using light bulbs are 250 times faster than broadband," *Independent*, 2013. [Online]. Available: http://www.independent.co.uk/news/science/li-fi-revolution-internet-connections-using-light-bulbs-are-250-times-faster-than-broadband-8909320.html.

25 D.-R. Kim, S.-H. Yang, H.-S. Kim, Y.-H. Son, and S.-K. Han, "Outdoor visible light communication for inter-vehicle communication using controller area network," *2012 Fourth International Conference on Communications and Electronics (ICCE)*, pp. 31–34, 2012.

26 X. Ng and W. Chung, "VLC-based medical healthcare information system," *Biomed. Eng. Appl. Basis Commun.*, vol. 24, no. 2, pp. 155–163, 2012.

27 R. Murai et al., "A novel visible light communication system for enhanced control of autonomous delivery robots in a hospital," *2012 IEEE/SICE International Symposium on System Integration*, vol. 3000, pp. 510–516, 2012.

28 Park, S. B., Jung, D. K., Shin, H. S., Shin, D. J., Hyun, Y. J., Lee, K., & Oh, Y. J. (2007). Information broadcasting system based on visible light signboard. *Proc. Wireless Opt. Commun*, 30, 311–313.

29 Y. Wang, N. Chi, Y. Wang, L. Tao, and J. Shi, "Network architecture of a high-speed visible light communication local area network," *IEEE Photonics Technol. Lett.*, vol. 27, no. 2, pp. 197–200, 2015.

30 T. Matsumura, "10 Mbps visible light transmission system," *Visible Light Communications Consortium.* [Online]. Available: https://mentor.ieee.org/802.15/dcn/08/15-08-0171-00-0vlc-10mbps-visible-lighttransmission-system-system.pdf.

31 R. Ferreira, E. Xie, J. Mckendry, and S. Rajbhandari, "High bandwidth GaN-based micro-LEDs for multi-Gbps visible light communications," *IEEE Photonics Technol. Lett.*, vol. 28, no. 19, p. 1, 2016.

32 S. Shao et al., "Design of a visible-light-communication enhanced WiFi system," pp. 1–10, 2015.

33 M. Ayyash et al., "Coexistence of WiFi and LiFi towards 5G."

34 M. Kodama and S. Haruyama, "Pulse width modulated visible light communication using digital micro-mirror device projector for voice information guidance system," *2019 IEEE 9th Annual Computing and Communication Workshop and Conference (CCWC)*, pp. 811–816, 2019.

35 G. Ntogari, T. Kamalakis, J. W. Walewski, and T. Sphicopoulos, "Combining illumination dimming based on pulse-width modulation with visible-light communications based on discrete multitone," *J. Opt. Commun. Netw.*, vol. 3, no. 1, p. 56, 2010.

36 A. Pradana, N. Ahmadi, and T. Adionos, "Design and implementation of visible light communication system using pulse width modulation," *5th International Conference on Electrical Engineering and Informatics (ICEEI 2015)*, pp. 25–30, 2015.

37 M. H. Ahfayd, M. J. N. Sibley, P. J. Mather, and P. I. Lazaridis, "Visible light communication based on offset pulse position modulation (offset-PPM) using high power LED," *2017 32nd General Assembly and Scientific Symposium of the International Union of Radio Science URSI GASS 2017*, vol. 2017-January, no. August, pp. 1–4, 2017.

38 A. Y. Ali, Z. Zhang, and B. Zong, "Pulse position and shape modulation for visible light communication system," *2014 International Conference on Electromagnetics in Advanced Applications (ICEAA)*, pp. 546–549, 2014.

3

Radio over Fiber System

3.1 Overview

In the early 1980s, in the USA, RoF was developed to use as a moving control station (CS) away from radar emitter missiles [1]. Cooper was the first to use RoF technology for mobile communication [2]. Radio over fiber technology has contributed to almost all fields, from military to civilian, and in all areas. It provides a reliable transmission medium and connectivity solutions to current and future high-speed indoor/outdoor communication, and it has the capability to support indoor fiber to the home (FTTH), second generation (2G), third generation (3G), long-term evolution (LTE), fourth generation (4G), and upcoming fifth generation (5G) and sixth generation (6G) networks. In this chapter, several RoF system designs and analysis work are presented.

Sending radio signals over an optical medium is known as RoF technology [3–12]. Simplicity, low power, low attenuation loss, and high bandwidth are some of the main benefits of RoF technology [13–17]. The base station's primary task is to convert an optical signal to a radio signal and vice versa. Figure 3.1a illustrates a simple, bidirectional RoF system. After some advancement, the photodiode and laser drivers were placed with a single unit called an electro-absorption modulator (EAM) [18], as depicted in Figure 3.1b. The optical carrier frequency is 200 THz, whereas in microwaves it is 1 GHz. So, due to a high-frequency carrier, there is an increase in data transmission and, theoretically, unlimited bandwidth. Technologies such as wavelength division multiplexing (WDM) and optical time domain multiplexing (OTDM) increase the bandwidth of an optical system [19, 20].

Radio over fiber is an analog transmission system. The analog signal, which can be a radio frequency (RF), an intermediate frequency (IF), or a baseband (BB) signal, is transmitted over optical fiber. At the transmitter, the RF/IF/BB signal is imposed by a direct modulation or an external modulation technique.

Hybrid Communication Systems for Future 6G and Beyond:
Visible Light Communication & Radio over Fiber Technology, First Edition. Rao Kashif.

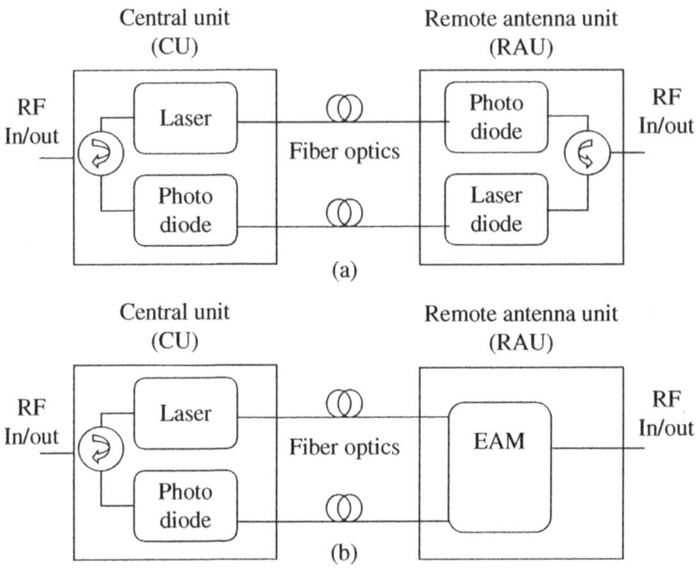

Figure 3.1 Radio over fiber base station (a) without EAM (b) with EAM.

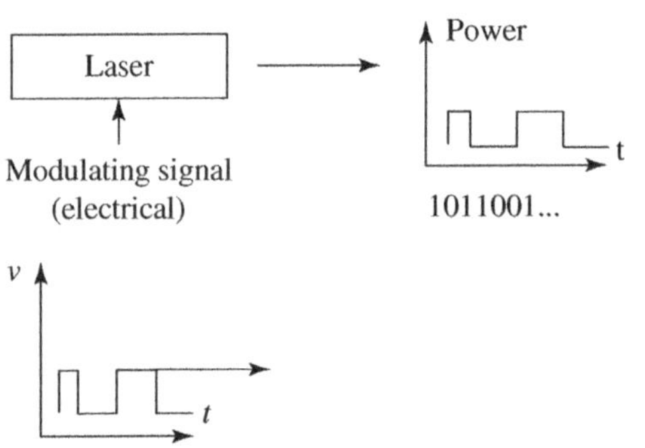

Figure 3.2 RoF direct modulation.

3.1.1 Direct Modulation

Figure 3.2' illustrates direct modulation in which the input drive current controls the output. The number 1 represents the emitting of light, while 0 represents the absence of light. This is the most straightforward method of modulation; however, it is slower and can be applicable for frequencies below 3 GHz.

3.1.2 External Modulation

In external modulation, the external device is used for the intensity/phase modulation of light. Modulators such as Linbo3, Mach-Zehnder (MZM), or an EAM are

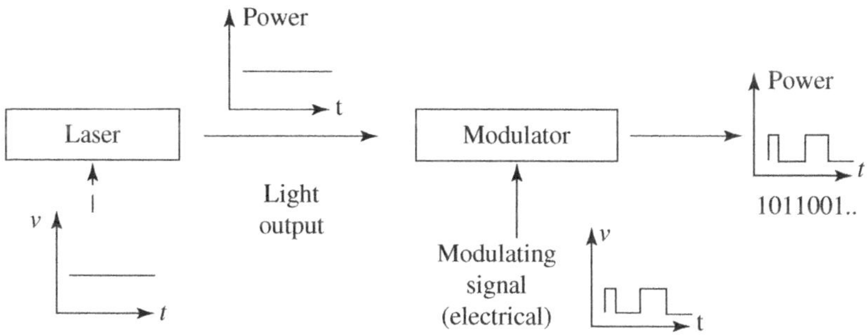

Figure 3.3 Radio over fiber externa modulation.

used as information shutters. Although this is an expensive method, it is faster and can be used for long-haul and high-power, solid-state lasers. Figure 3.3' illustrates external modulation.

3.2 Radio over Fiber Link Configuration

The RoF link configuration is based on the type of frequency bands such as BB, IF, and RF bands, as demonstrated in Figure 3.4.

3.2.1 Radio Frequency over Fiber

As illustrated in Figure 3.4a, the CS generates the modulated signal and transmits it to the base station by an electro-optical modulator (EOM). A photo detector detects the modulator signal and transmits it to mobile hosts [21].

It has a simple base station design; however, transmission distance is limited due to chromatic dispersion [22].

3.2.2 Intermediate Frequency over Fiber

Figure 3.4b depicts an IF-over-fiber link. In an IF band, the CS transmits the modulated signal to the base station by an EOM [23]. A photodiode then detects the transmitted signal, and after up-conversion to an RF band, it is transmitted to mobile hosts. Chromatic dispersion is less; however, additional equipment, such as millimeter (mm)-wave frequency LO frequency up- and down-conversion, is needed on the base station [24].

3.2.3 Baseband over Fiber

Figure 3.4c depicts a BB-over-fiber link. In BB over fiber, the CS generates a modulated signal in BB and transmits it to BB via an EOM. A photodiode then detects

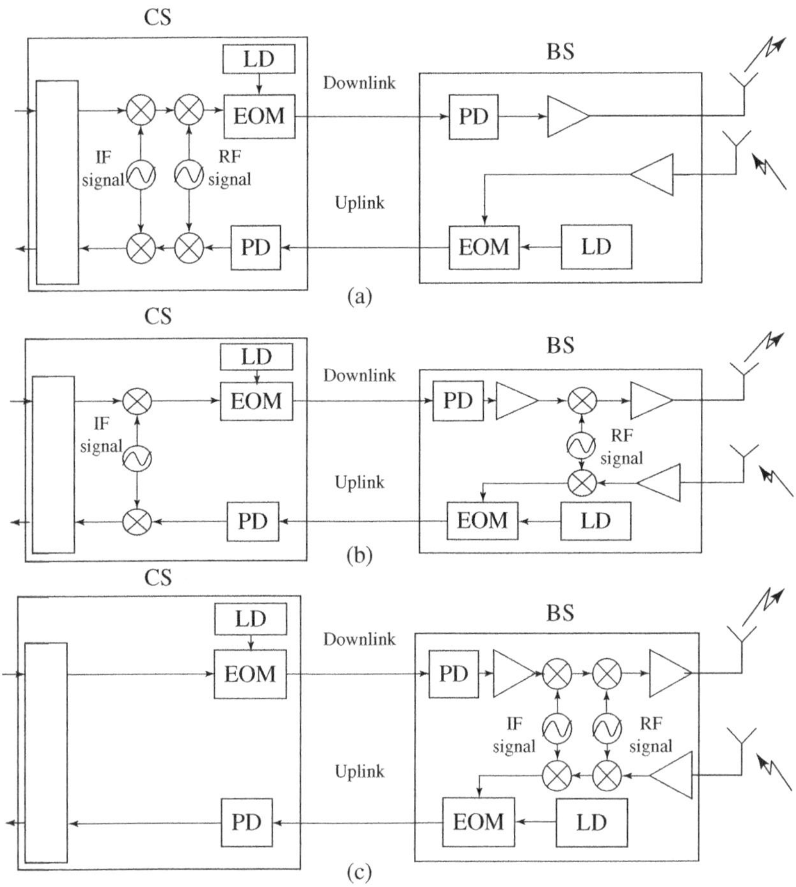

Figure 3.4 RoF link configuration (a) RF over fiber (b) IF over fiber (c) BB over fiber.

the modulated signal on the receiver; then, either after up-conversion to an RF band through IF or directly, it is transmitted to mobile hosts [25]. Fiber dispersion is negligible; however, BS has a complex structure.

The uplink from the mobile host to the CS is an inverse process. In RF over fiber, the base station sends the received signal to CS after it is amplified by LD using EOM. In IF and BB, the received signals are down-converted before being sent to the control center by LD using EOM [26].

3.2.4 Millimeter-Wave Signal Generation

Current research is being conducted in the field of mm-wave signal generation. Several techniques have been introduced [27]; among which optical heterodyning

Table 3.1 Millimeter-wave signal generation techniques.

Techniques	RF band (GHz)	IF band (GHz)	Bit rate (Mbps)	Modulation	Fiber length (km)	Refs.
Optical heterodyning	64 60	— —	155 155.52	OQPSK DPSK	12.8 25	[24] [25]
External modulation	59.6 60, 59.6	2.6 3.0, 2.6	156 156	DPSK DPSK	50 85	[26] [27]
Up and down conversion	40 40	3.0 3.0	156 155(down) 51.8 (up)	DPSK BPSK	50 40	[34] [46]
Optical transceiver	60 60, 59.6	— 3.0, 2.6	120 155.52	QPSK DPSK	13 27.5	[35, 36] [37]

[28–33], external modulation [33–36], up and down conversion [37–41], and optical transceivers [42–45] are famous. Table 3.1: A summary of findings from multiple reviews, as referenced in source [17].

3.2.5 Applications

3.2.5.1 Satellite Communication

To work efficiently, the satellite ground stations need high bandwidth carriers between the CS and the center office (CO). Optical fiber could be a prime candidate for providing connectivity between the CS and the CO. Future satellite broadband networks also require strong backhaul connectivity, such as optical fiber. Therefore, RoF has large applications in satellite-related services.

3.2.5.2 Cellular Networks

Due to the increase in services provided by cellular companies, the need for capacity is increasing, and with the revolution of 4G and the upcoming 5G, mobile network service providers require a high bandwidth channel for their networks. Here, RoF is the best option to support future cellular networks.

3.2.5.3 Transportation and Vehicles

The medium of transportation is becoming faster with the passing of time. High-speed trains and high-speed vehicles on highways all need high-frequency signals, and RoF can provide high bandwidth and high-frequency signal carrier channels between base stations mounted along highways or railway tracks and the CS.

3.2.5.4 Visible Light Communication

Visible light communication/Li-Fi is an upcoming new technology for indoor high-speed wireless communication. The practical speed of indoor VLC/Li-Fi is 10 Gb/s. To hold such large amounts of data, VLC requires a strong backhaul system to hold the data traffic, and RoF is the only solution to this problem.

3.3 Radio over Fiber System-Level Analysis

With the increasing demand for radio over technology in current and future communication, a system-level analysis is needed. All the components should be thoroughly analyzed and the signal integrity in RoF systems should be increased. In this section, we performed system-level analyses with different modulation techniques, and we checked system performance with positive-intrinsic negative (PIN) and avalanche photodiode (APD) photodiodes.

3.3.1 Encoding Formats

In optical communication, the electrical signal is converted to a bit stream by different encoding formats [45]. In non-return zero (NRZ), the number 1 represents a high amplitude, while 0 stands for a low amplitude [47]. This format is simpler, and less bandwidth is required; however, there is the absence of a clock, and in long-bit sequences, there will be a problem in synchronization. In return to zero (RZ), all the bits are returned to zero [47]. The half of 1 bit will remain high, but it soon returns to zero. Therefore, it occupies twice the bandwidth, compared to NRZ. In this experiment, we use two popular modulation formats, namely NRZ and RZ.

A pseudo-random binary sequence (PRBS) with a bit rate of 10 Gb/s is used for simulation, as illustrated in Figures 3.5, 3.6 and 3.7 present the oscilloscope and spectrum analysis of NRZ and RZ.

3.3.2 PIN and APD Photodiodes

A photodiode structure is a PN junction, so the photons that are incidental to the active region create electron-hole pairs. In this way, the electron-hole conduction across the depletion layer gives rise to a photocurrent. Every absorbed photon creates an electron-hole pair. In section 3.4 comparison analysis of both PIN and APD is presented.

3.4 Simulation

Figure 3.8' depicts a simple RoF that we designed for analysis in the OptiSystem 14 tool. The bit rate was set to 10 Gb/s, and on the receiver side, both the APD and PIN photodiodes were used to check system performance.

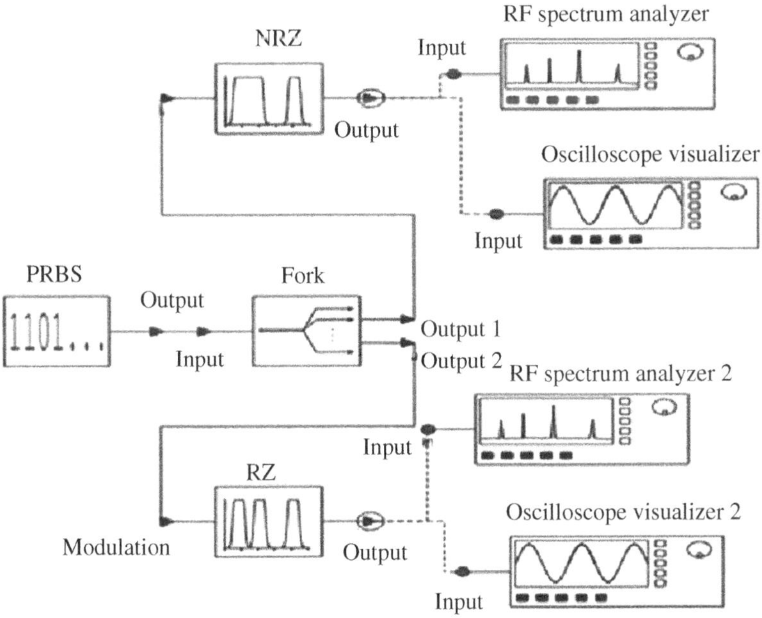

Figure 3.5 NRZ and RZ simulation setup.

The length of optical fiber was kept to 5, 10, and 15 km. At 5 km, the *Q*-factor of the PIN photodiode was high, compared to that of the APD. However, with the increase of distance from 10 to 15 km, the APD *Q*-factor began to increase compared to PIN. Figure 3.9 presents a *Q*-factor graph of PIN and APD, while Figure 3.10 depicts an eye diagram of both at 15 km.

3.4.1 Result

NRZ and RZ were analyzed at a 10 Gb/s bit rate, and NRZ was found to be more efficient because of less bandwidth usage. As the PIN and APD photo detectors, the *Q*-factors were compared, and PIN demonstrated good performance at shorter distances, while the APD *Q*-factor was higher at far distances. Therefore, we can say that systems designed for short-distance communication, such as LAN, can use PIN photodiodes, and for wide-area networks and long-distance communication, APD can be a smart choice. APD is a highly sensitive semiconductor device.

3.5 Future Multifunctional RoF Home Network

In the future, with the implementation of the IoT, wireless systems will handle multifrequency applications on the same medium of transmission. Radio over

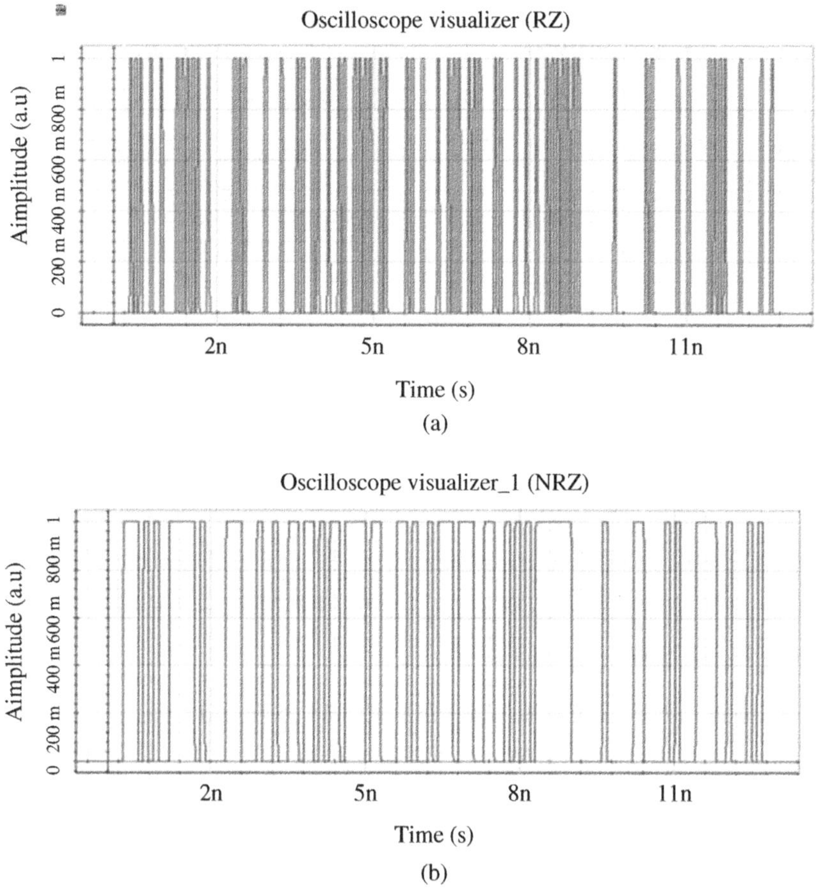

Figure 3.6 Oscilloscope analysis (a) NRZ (b) RZ.

fiber can serve well in this situation, as it can handle multifrequency channels on a single medium, as demonstrated in Figure 3.11.

3.5.1 Fiber to the Home (FTTH)

FTTH is the method of transporting signals from the point of origin, which might be called a head or hub, out to a subscriber. It is also a solution that does not require any active devices; therefore, it allows us to deliver content at a greatly reduced operating expense as well as over longer distances and for higher volumes of data [38–43]. The largest drivers are certainly speed and the capacity of the fiber. Fiber can transport data across large distances and generally does not require any amplification to reach the service area, and this leads to a reduction

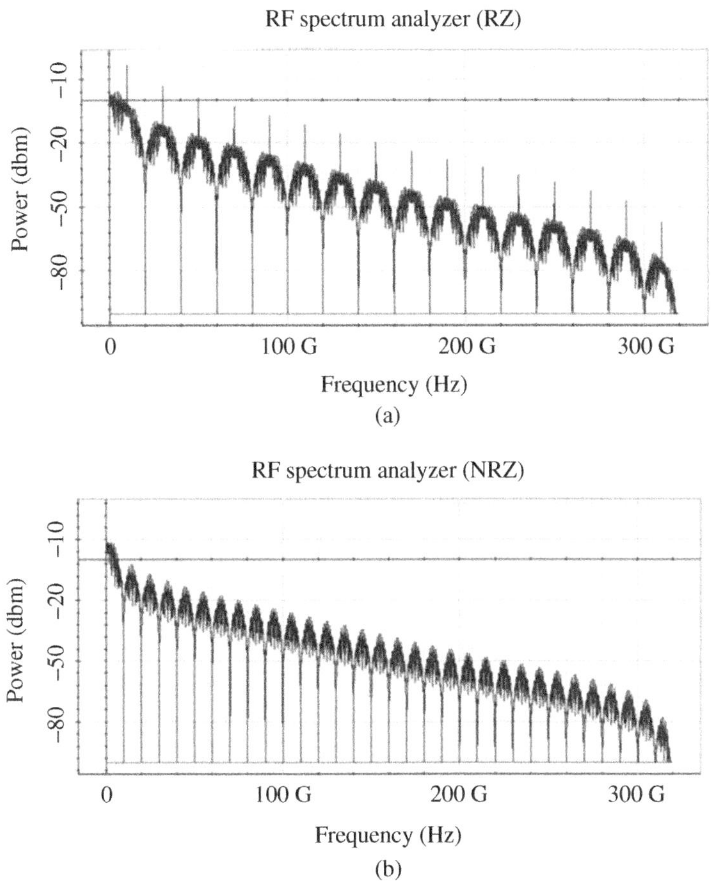

Figure 3.7 Spectrum analysis (a) NRZ (b) RZ.

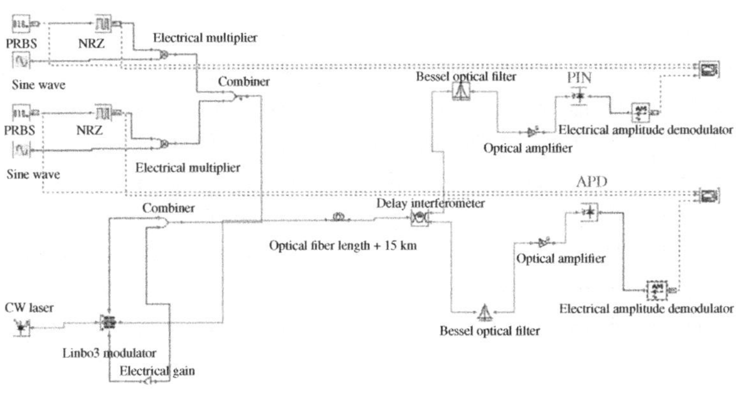

Figure 3.8 Radio over fiber system in OptiSystem.

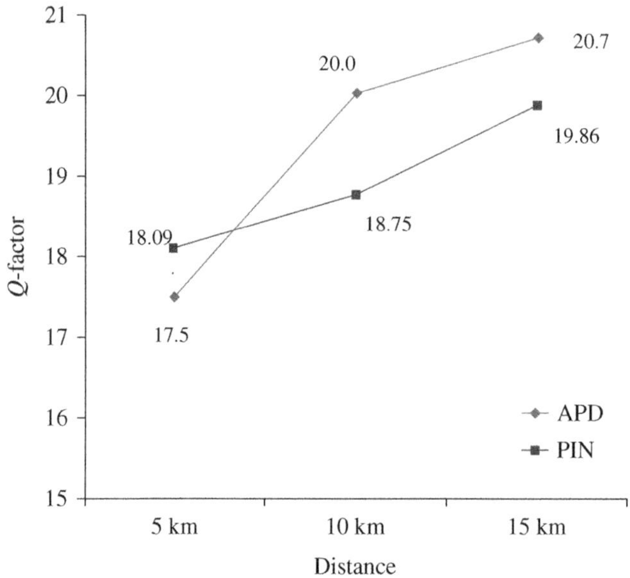

Figure 3.9 PIN & APD Q-factor graph.

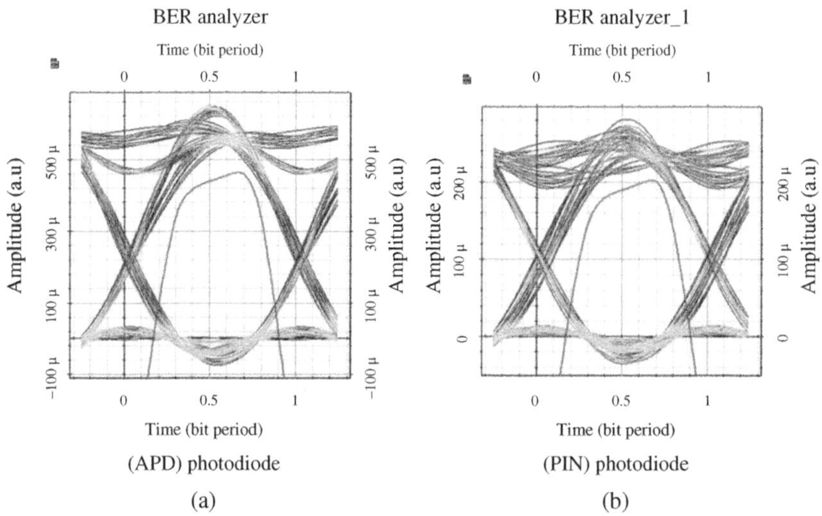

Figure 3.10 PIN & APD eye diagram.

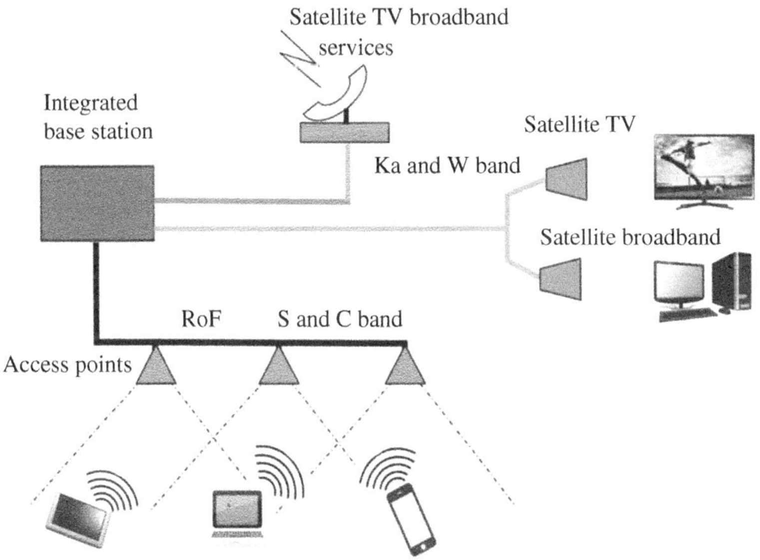

Figure 3.11 Multifrequency RoF system.

of many devices that require maintenance and consume much power. Moreover, fiber delivers much higher bandwidths than any traditional alternatives.

3.5.2 Multifrequency RoF System Design

Future home networks will be the integration of different frequency applications, and RoF will support FTTH networks, as demonstrated in the Figure 3.11.

Therefore, we performed a system-level simulation of RoF over 10 km with 2.5 and 29 GHz signals on a single fiber. Figure 3.12 depicts a block diagram, and the

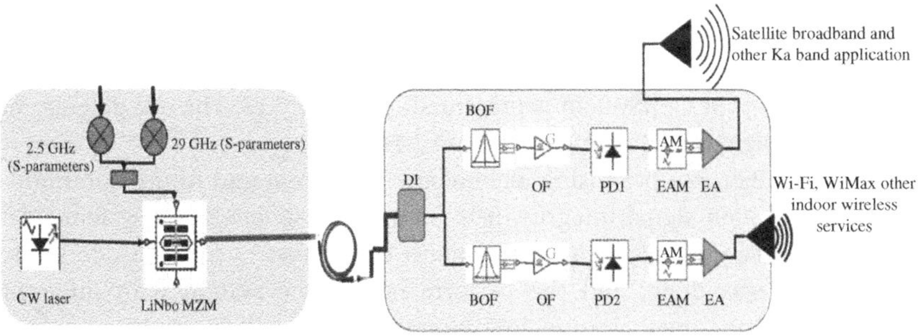

Figure 3.12 Experiment RoF system overview.

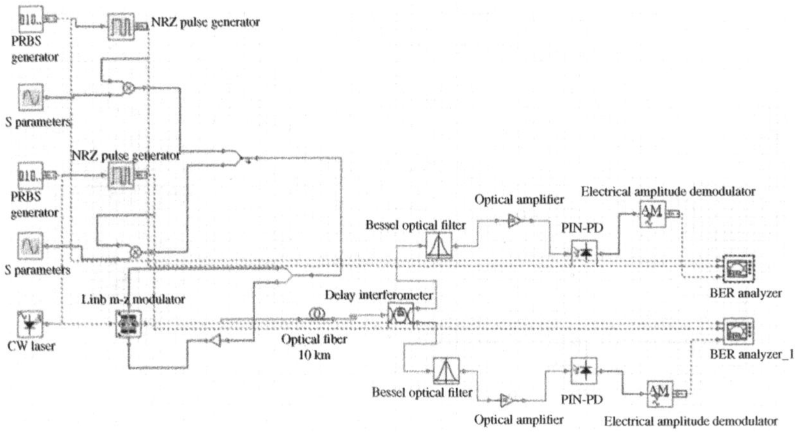

Figure 3.13 Multifrequency RoF system in OptiSystem.

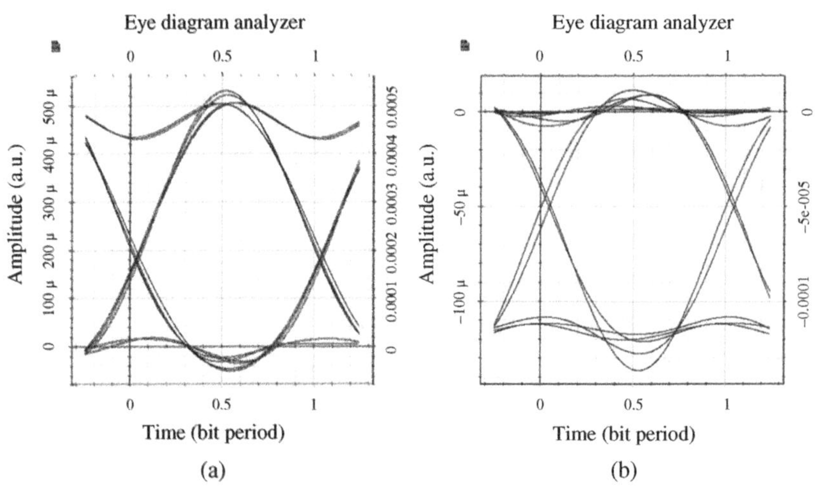

Figure 3.14 Eye diagram for (a) 2.5 GHz and (b) 29 GHz.

simulated system in OptiSystem is presented in Figure 3.13. The eye diagram of both received signals is illustrated in Figure 3.14.

Radio over fiber is a promising technology for current and future communication. To maintain signal integrity, it is compulsory to conduct a system-level performance analysis, taking all components of the system into consideration. This analysis was done, and the performance of the system with different photodiodes is presented. Future IoT communication will be multi-frequency communication, and RoF can handle multiple frequencies on a single medium. The system-level analysis was conducted with 2.5 and 5 GHz.

References

1 S. Komaki, *Radio Over Fiber Technologies for Mobile Communications Networks*. Artech House, 2002.
2 A. J. Cooper, "Fibre/radio' for the provision of cordless/mobile telephony services in the access network," *Electron. Lett.*, vol. 26, no. 24, pp. 2054–2056, 1990.
3 C. Lim, A. Nirmalathas, Y. Yang, D. Novak, and R. Waterhouse, "Radio-over-fiber systems," *Commun. Photonics Conf. Exhib. (ACP)*, vol. 7632, pp. 1–10, 2009.
4 J. J. Vegas Olmos and I. Tafur Monroy, "Reconfigurable radio-over-fiber networks," in *[OFC] Optical Fiber Communication Conference*, p. W1F.3, 2015.
5 A. M. J. Koonen *et al.*, "Perspectives of radio over fiber technologies," in *OFC/NFOEC 2008 - 2008 Conference on Optical Fiber Communication/National Fiber Optic Engineers Conference*, 2008.
6 H. Al-raweshidy and S. Komaki, "Radio over fiber technology for the next generation," *Radio Over Fiber Technol. Mob. Commun. Networks*, Vol. 1 pp. 183–199, 2002.
7 N. J. Gomes and D. Wake, "Introduction to radio over fiber," in Nathan J. Gomes, Paulo P. Monteiro, Atílio Gameiro *Next Generation Wireless Communications Using Radio over Fiber*, Wiley. pp. 61–89, 2012, https://onlinelibrary.wiley.com/doi/abs/10.1002/9781118306017.ch4#.
8 D. Novak *et al.*, "Radio-over-fiber technologies for emerging wireless systems," *IEEE J. Quantum Electron.*, vol. 52, no. 1, pp. 1–11, 2016.
9 D. Opatic, "Radio over Fiber Technology for Wireless Access," *Mipro*, 2009.
10 S. Abdollahi, H. Al-Raweshidy, A. Ahmadinia, and R. Nilavalan, "An all-photonic digital radio over fiber architecture," in *2011 IEEE Swedish Communication Technologies Workshop, Swe-CTW 2011*, pp. 62–67, 2011.
11 D. Wake, M. Webster, G. Wimpenny, K. Beacham, and L. Crawford, "Radio over fiber for mobile communications," *2004 IEEE International Topical Meeting on Microwave Photonics(IEEE Cat. No.04EX859)*, pp. 157–160, 2004.
12 M. García Larrodé, A. Koonen, J. Vegas Olmos, and A. Ng'oma, "Bidirectional radio-over-fiber link employing optical frequency multiplication," *IEEE Photonics Technol. Lett.*, vol. 18, no. 1, pp. 241–243, 2006.
13 R. Kashif, O. Famoriji, and F. Lin, "Radio over fiber system level performance analysis, maintaining signal integrity," *J. Comput. Commun.*, vol. 5, no. 13, pp. 1–8, 2017.
14 M. Lach, E. Schuh, and K. Schmidt, "Application of electroabsorption modulators for high-speed transmission," in *Ultrahigh-Speed Optical Transmission Technology*, Journal of Optical and Fiber Communications Reports, Berlin: Springer, 2005, pp. 347–377.

15 R. Karthikeyan and S. Prakasam, "A survey on radio over fiber (RoF) for wireless broadband access technologies," *Int. J. Comput. Appl.*, vol. 64, no. 12, 2013.

16 Y. L. Y. Lin and P. T. P. Tien, "Next-generation OFDMA-based passive optical network architecture supporting radio-over-fiber," *IEEE J. Sel. Areas Commun.*, vol. 28, no. 6, pp. 791–799, 2010.

17 V. Vorgelegt, "Radio over fiber based network architecture," *PhD. Dissertation , Department of Electrical Engineering, Berlin University, Berlin*, 2005.

18 T. N. N. U. Gliese, S. Norskow, "Chromatic dispersion in fiber-optic microwave and millimeter-wave links," *IEEE Trans. Microw. Theory Tech*, vol. 44, no. 10, pp. 1716–1724, 1996.

19 K. Kitayama, "Architectural considerations of radio-on-fiber millimeter-wave wireless access systems, Signals, Systems, and Electron," in *URSI International Symposium*, pp. 378–383, 1998.

20 R. H. R. Hofstetter, H. Schmuck, "Dispersion effects in optical millimeter-wave systems using self-heterodyne method for transport and generation," *IEEE Trans. Microw. Theory Tech*, vol. 43, no. 9, pp. 2263–2269, 1995.

21 R. H. J. J. O'Reilly, P. M. Lane, R. Heidemann, "Optical generation of very narrow linewidth millimeter wave signals," *Electron. Lett*, vol. 28, pp. 2309–2311, 1995.

22 R. S. T. D. Novak, Z. Ahmed, R. B. Waterhouse, "Signal generation using pulsed semiconductor lasers for application in millimeter-wave wireless links," *IEEE Trans. Microw. Theory Tech*, vol. 43, pp. 2257–2262, 1995.

23 M. Okuno, K. Kitayama, T. Kuri, H. Yokoyama, "60 GHz millimeter-wave generation and transport using stabilized mode-locked laser diode with optical frequency DEMUX switch," in *Proceedings of GLOBECOM'96. 1996 IEEE Global Telecommunications Conference*, pp. 2126–2169, 1996.

24 F. S. R. Braun, G. Grosskopf, D. Rohde, "Low-phase-noise millimeter-wave generation at 64 GHz and data transmission using optical sideband injection locking," *IEEE Photon. Technol. Lett*, vol. 10, no. 5, pp. 728–730, 1998.

25 K. Kuri, "Optical heterodyne detection technique for densely multiplexed millimeter-wave-band radio-on-fiber systems," *J. Light. Technol*, vol. 21, no. 12, pp. 3167–3179, 2003.

26 Y. Kuri, K. Kitayama, "Fiber-optic millimeter-wave uplink system incorporating remotely fed 60-GHz-band optical pilot tone," *IEEE Trans. Microw. Theory Tech*, vol. 47, no. 7, pp. 1332–1337, 1999.

27 Y. Kuri, K. Kitayama, A. Stöhr, "Fiber-optic millimeter-wave downlink system using 60 GHz-band external modulation," *J. Light. Technol*, vol. 17, no. 5, pp. 799–806, 1999.

28 Z. Smith, D. Nowak, "Technique for optical SSB generation to overcome dispersion penalties in fibre-radio systems," *Electron. Lett*, vol. 33, no. 1, pp. 74–75, 1997.

29 K. Park, W. Sorin, "Elimination of the fibre chromatic dispersion penalty on 1550 nm millimeter-wave optical transmission," *Electron. Lett*, vol. 33, no. 6, pp. 512–513, 1997.

30 R. Williams, "Optically amplified downconverting link with shot-noise-limited performance," *IEEE Photon. Technol. Lett*, vol. 8, no. 1, pp. 148–150, 1996.

31 S. Sun, R. Orazi, "Efficient microwave frequency conversion using photonic link signal mixing," *IEEE Photon. Technol. Lett*, vol. 8, no. 1, pp. 154–156, 1996.

32 C. Helkey, J. Twichell, "A down-conversion optical link with RF gain," *J. Light. wave Technol*, vol. 15, no. 6, pp. 956–961, 1997.

33 G. Smith and D. Novak, "Broadband millimeter-wave fiber-radio network incorporating remote up/down conversion," in *Microwave Symposium Digest*, pp. 1509–1512, 1998.

34 K. Kitayama and R. Griffin, "Optical downconversion from millimeter-wave to IF-band over 50-km-long optical fiber link using an electroabsorption modulator," *IEEE Photon. Technol. Lett*, vol. 11, no. 2, pp. 287–289, 1999.

35 D. Nöel, D. Wake, D. Moodie, D. Marcenac, and L. Westbrook, "Novel techniques for high-capacity 60-GHz fiber-radio transmission systems," *IEEE Trans. Microw. Theory Tech*, vol. 45, no. 8, pp. 1416–1423, 1997.

36 D. Ager, A. Stöhr, and Y. Kitayama, "Full-duplex fiber-optic RF subcarrier transmission using a dual-function modulator/photodetector," *IEEE Trans. Microw. Theory Tech*, vol. 47, no. 7, pp. 1338–1341, 1999.

37 D. Ager, Y. Kitayama, A. Stöhr, T. Kuri, and R. Heinzelmann, "An approach to single optical component antenna base stations for broad-band millimeter-wave fiber-radio access systems," *IEEE Trans. Microw. Theory Tech*, vol. 48, pp. 2588–2594, 2000.

38 Y. Kuri, K. Kitayama, "60-GHz-band full-duplex radio-on-fiber system using two-RF-port electroabsorption transceiver," *IEEE Photon. Technol. Lett*, vol. 12, no. 4, pp. 419–421, 2000.

39 Y. Kitayama, K. Ikeda, T. Kuri, A. Stöhr, "Full-duplex demonstration of single electroabsorption transceiver basestation for mm-wave fiber-radio systems," in *International Topical Meeting on Microwave Photonics, MWP 2001/MWP 2001*, vol. 2001, pp. 73–76, 2001.

40 Maxim Integrated, "NRZ bandwidth – HF cutoff vs. SNR," pp. 2–6, 2008.

41 X. Liang, W. Li, N. Linze, "Comparison of return to zero and non-return-to zero coded pulses for BOTDA," in *The 9th International Conference on Optical Communication and Networks (ICOCN 2010)*, 2010.

42 S. Cherry, "Fiber to the home," *IEEE Spectrum*, vol. 41, no. 1. pp. 42–43, 2004.

43 T. Koonen, "Fiber to the home/fiber to the premises: What, where, and when?," *Proc. IEEE*, vol. 94, no. 5, pp. 911–934, 2006.

44 P. Shumate, "Fiber-to-the-home: 1977–2007," *J. Light. Technol.*, vol. 26, no. 9, pp. 1093–1103, 2008.

45 C. Lee, W. Sorin, and B. Kim, "Fiber to the home using a PON infrastructure," *J. Light. Technol.*, vol. 24, no. 12, pp. 4568–4583, 2006.

46 G. Smith and D. Novak, "Broadband millimeter-wave fiber-radio network incorporating remote up/down conversion," in *Microwave Symposium Digest*, 1998, pp. 1509–1512.

47 C. Lee and K. Choi, "Fiber to the home," in *Conference Proceedings – Lasers and Electro-Optics Society Annual Meeting-LEOS*, pp. 937–938, 2007.

4

Digital Coherent Integration with Radio over Fiber

4.1 Digital Coherent System Analysis

The digital coherent system is a promising technology for future high-speed systems [1]. It has been the favorite technology for long-haul optical designers, and much research has been done to improve its signal integrity [1–13]. After the implementation of 100-GB systems, there is now much research and development work on 400-GB next-generation systems [13–25]. Figure 4.1 illustrates the commercial deployment of the map [1].

A digital coherent optical system can be categorized into five parts: transmitter, channel, receiver, digital signal processing (DSP), and decision/polarization. Figure 4.2 presents an overview of a digital coherent optical transceiver system. As is evident, the system is based on five main portions.

To gain a better understanding of optical and electrical signals, the former are highlighted in blue, and the latter is highlighted in grey color.

Figure 4.3 provides a clearer understanding of signal information. We can see that the transmitter's end portion and the receiver's front will always be in contact with optical signals, and the work done on this portion will primarily be related to, inter alia, optoelectronic circuits, photo detectors, and silica couplers [26]. The front of the transmitter and the end of the coherent receiver are related to electrical signals; therefore, all DSP and high-speed electronic circuit-related work will be done in these two areas [27, 28]. The transmitter sends a signal to the digital coherent receiver via the optical channel, and after the signal is received in light form, it is mixed with the local light oscillator. Photo detectors will pass over information to the ADC, which will be further processed by the DSP and decision portion.

Hybrid Communication Systems for Future 6G and Beyond:
Visible Light Communication & Radio over Fiber Technology, First Edition. Rao Kashif.

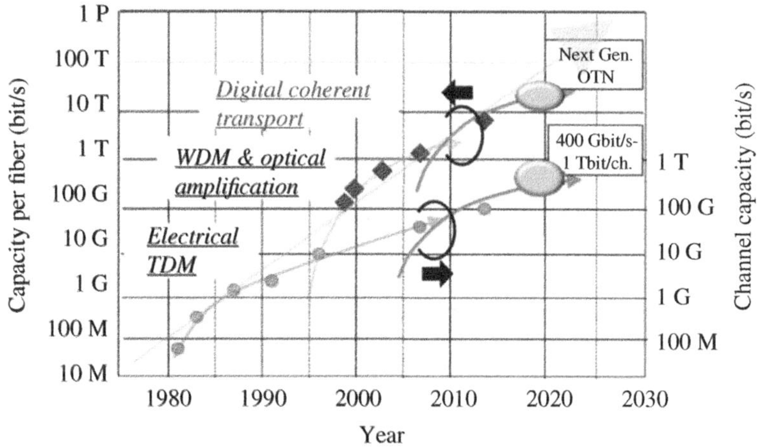

Figure 4.1 Optical commercial development map.

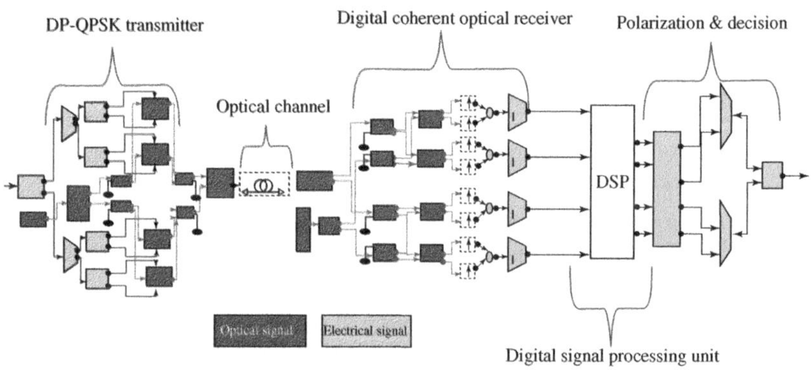

Figure 4.2 Digital coherent optical transceiver system overview.

4.1.1 DP-QPSK Transmitter

The transmitter is the first part of a digital coherent optical transceiver system, and it transmits the optical signal to the receiver through an optical channel. The DSP, optical modulator, modulator drivers, and light source are all parts of the transmitter [29]. Figure 4.4 provides an overview of a subsystem diagram of a transmitter.

The transmitter has both optical and electrical components – the color blue indicates optical signals and are related to optoelectronic work with silica circuits, while grey represents electrical signals that involve DSP on the front end before the signal reaches the optical modulator. The modulator is an important

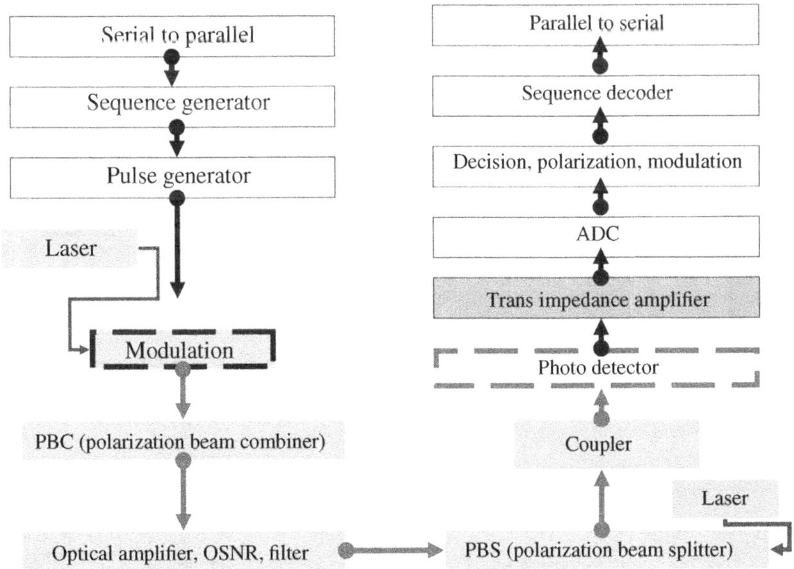

Figure 4.3 Optical and electrical signal flow in digital coherent optical transceiver.

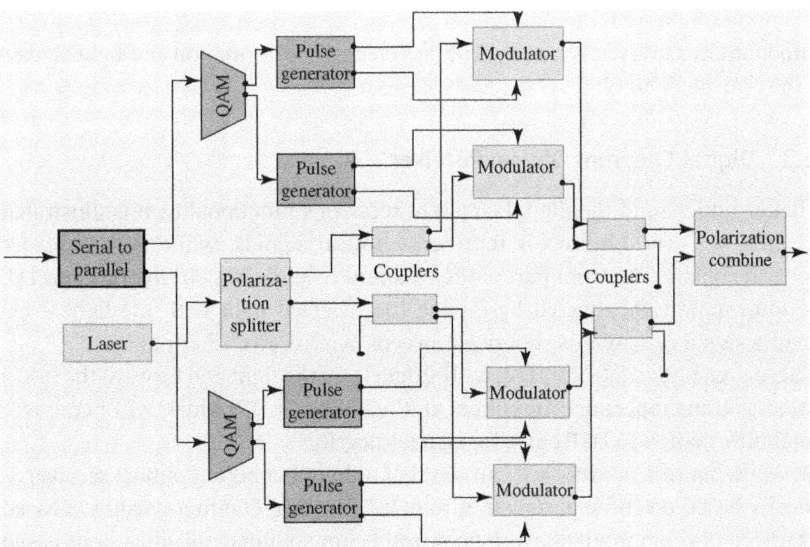

Figure 4.4 Digital coherent transmitter.

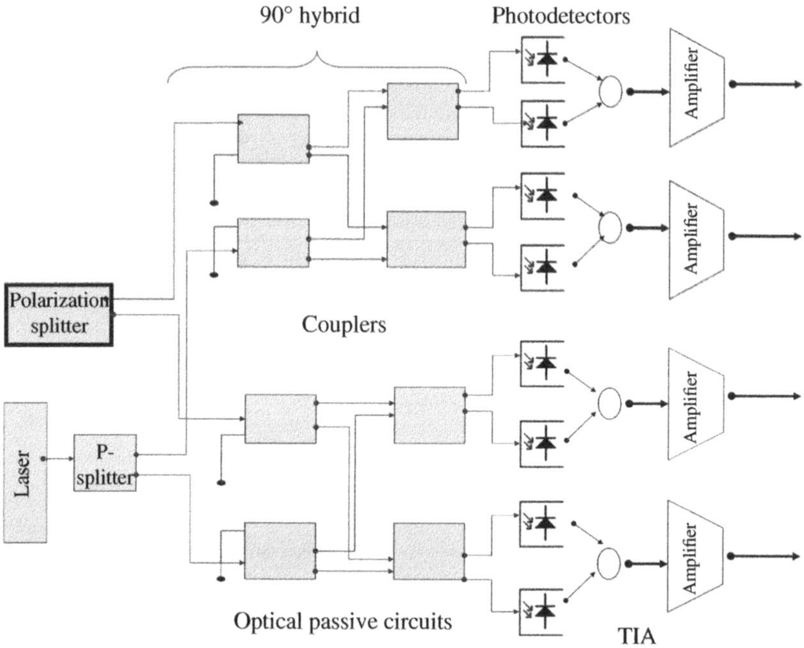

Figure 4.5 Digital coherent receiver.

component in a transceiver, as it combines electrical information and light signals for the optical channel.

4.1.2 Digital Coherent Optical Receiver

To better understand the digital coherent receiver's functionality, it is illustrated in Figure 4.5. The blue color represents optical signals, while grey indicates electrical signals. The front part of the coherent receiver has 90° hybrids and LO. The information is extracted by mixing the received light with LO light – LO provides signal gain and its shot noise sweeps over receiver thermal noise.

Thereafter, photo detectors receive the data from the light and forward the information to transimpedance amplifiers and further to ADS, which will hand over digital information to DSP chips for signal integrity.

By knowing this process, we can see that a digital coherent optical receiver is basically based on three parts, other than a DSP chip. The first portion is based on passive optical circuits having polarized beam splitters and silica light-based circuits.

The second part consists of high-speed optical semiconductors that are ultra-high-speed photo detectors, and the last part is made up of trans-impedance amplifiers, which are high-speed electronic devices. Since signal integrity is the main issue for all high-speed designs [30], it is necessary to work in all three fields to obtain better integration. The following are the challenges and tasks of these three parts of the receiver.

4.1.3 Optical Integration Technology

To implement a high-quality optical coherent receiver, it is necessary to integrate silica-based PLC technology, optical semiconductor technology, and high-speed electronic device technology, as demonstrated in Figure 4.5.

4.1.3.1 PLC Technology

To accurately receive digital coherent signals, it is necessary to detect the state of optical signals in a highly precise manner. If there are two different circuits, they usually have different performance characteristics.

Since this affects the performance of the entire system, the detection of signals tends to become less accurate. To solve this problem, new principles should be adopted, in which two matching circuits detect the phase information. Even if the two circuits have some deviation, it should be ensured that they deviate by the same amount.

4.1.3.2 Optical Semiconductor

In digital coherent transmission, it is necessary that the photodiode receives a stronger light than in conventional transmission methods. When there is stronger incident light, the response of electrons and holes conventionally becomes slower. The structure should be improved to keep response times short.

4.1.3.3 High-Speed Electronic Devices

The trans-impedance amplifier used for digital coherent transmission is required to offer a variety to provide, for example, automatic gain control manually and a shutdown capability. Also, to further improve transmission quality, the amplifier must keep its generated noise and distortion low, and it must exhibit a wide dynamic range. It takes a great deal to satisfy these requirements since low distortion and low noise are mutually conflicting. As a result, achieving both of these simultaneously in circuit design is extremely difficult, and an ingenious circuit structure must thus be developed to do so.

4.1.4 Digital Signal Processing in a Coherent Receiver

The advancement in ADC makes it possible to deploy a DSP module in a coherent receiver. The digital coherent technology combines coherent detection and DSP at the receiver end [31]. Figure 4.6 presents a DSP module schematic [31].

In the digital coherent receiver, the incoming data in light form will be mixed with LO light and then polarized. Electrical signals from photo detectors will be

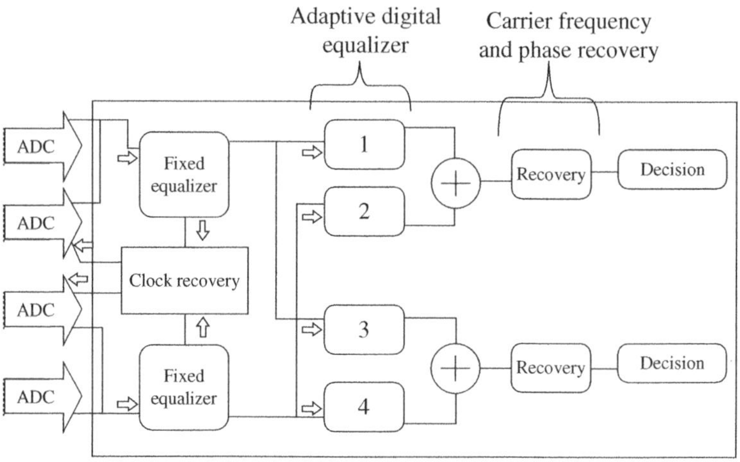

Figure 4.6 A DSP module in a digital coherent receiver.

digitalized by the ADC and sent to the DSP unit, which is the central part of the digital coherent receiver. Digital compensation is performed by two fixed equalizers at the front end, four adaptive digital equalizers, and phase recovery units at the end of the receiver.

Dual polarization QPSK sends two signals, separating horizontal and vertical polarization [32]. Both signals are polarized 90° from each other, and the signal could be 0/0, 1/1, 0/1, or 1/0, thereby yielding four bits on the receiver because optical signals are multiplexed into 4 bits per symbol in DP-QPSK after 2-bit multiplication in QPSK and further 2-bit multiplication in two orthogonal polarization states.

This is evident in Figure 4.7, in which a typical QPSK receiver schematic is presented. DP-QPSK is still not mature, and the need for ultra-fast DSP is the major challenge. DSP should meet the speed of the optical signal. According to Milorad Cvijetic [33], "Without DSP, we still face some issues, and I would say, it will not be quite feasible to apply these three great technologies". At present, DP-QPSK is the favorite; however, work on it is ongoing, and in the near future, there will be more techniques, as DSP technology is also heading toward fast circuits to meet the speed of optical signals.

4.2 Software Implementation

For a clearer and better understanding of the digital coherent optical system, the subsystems of a digital coherent transceiver are designed in OptiSystem and simulated at 115 Gb/s to generate constellation diagrams. In Figure 4.8, all subsystems are connected for simulation.

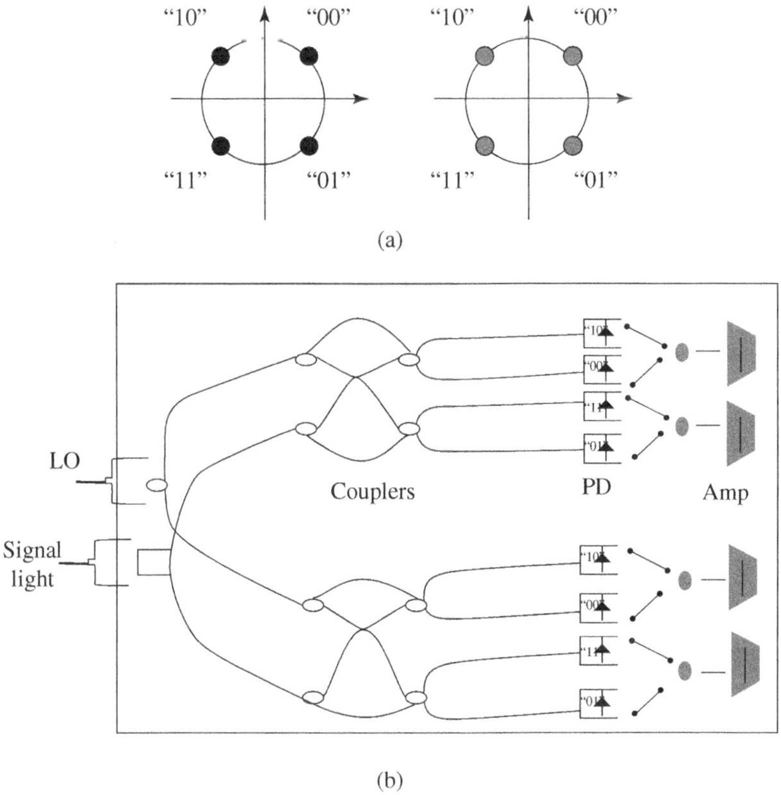

Figure 4.7 DP-QPSK (a) modulation (b) receiver schematic.

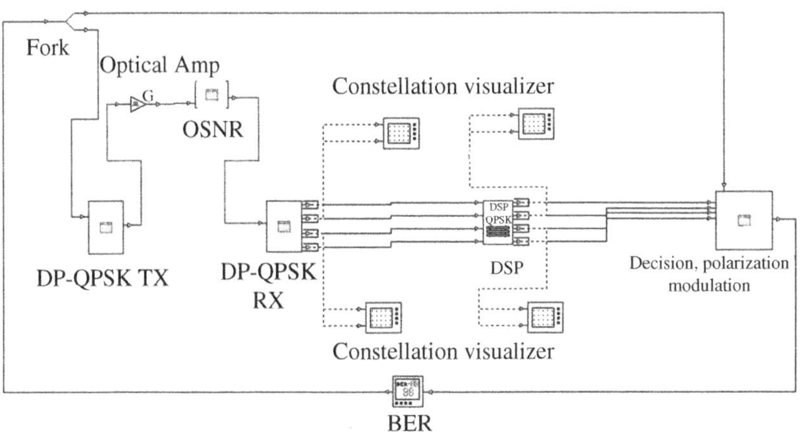

Figure 4.8 System design for 115 Gbps.

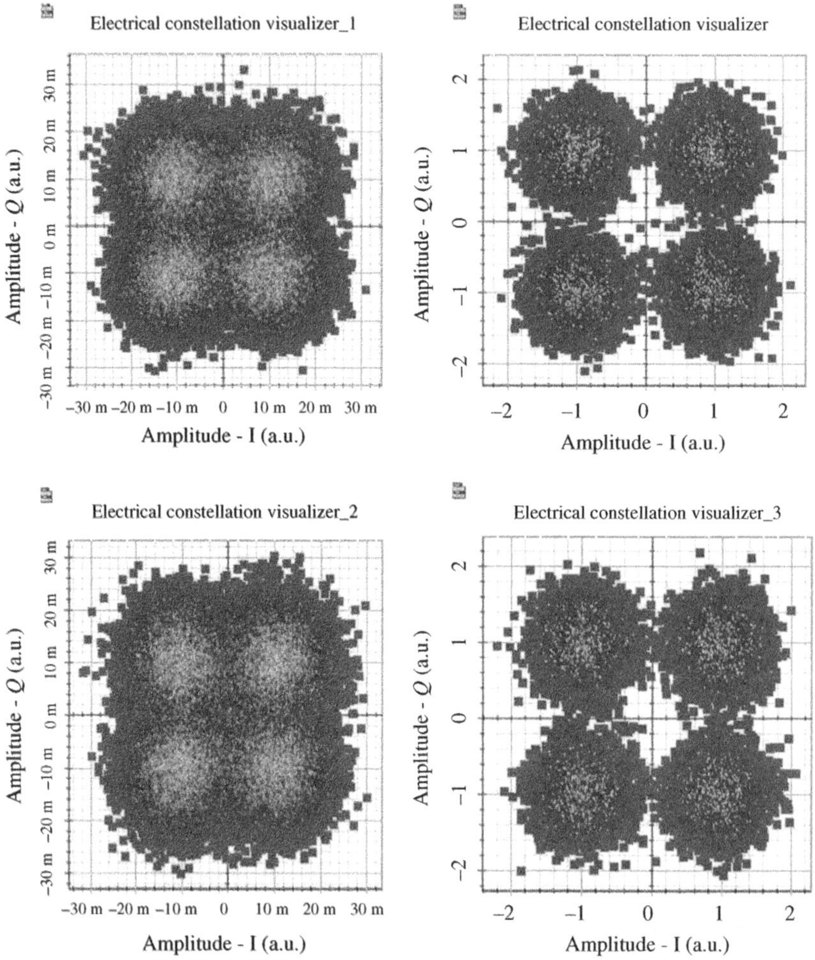

Figure 4.9 Constellation diagram (1, 3) before DSP (2, 4) after DSP.

The functions of all subsystems are already discussed in the upper section. Figure 4.9 depicts the constellation diagrams of generation 2 polarization*2 bit/symbol – successfully generated and compared before and after the DSP unit.

4.3 Digital Coherent RoF System Analysis

Visible light communication is one of the solutions to problems such as spectrum conjunction and the need for high-speed indoor communication. However, such a high-speed indoor system requires a powerful backhaul system that can handle this multi-gigabit data communication without facing any issues. A digital coherent is suitable for optical communication systems; therefore, we designed and simulated a digital coherent-based RoF system that can be a prime candidate for supporting backhaul communication for indoor VLC. Digital coherent systems can fulfill the high-capacity network needs since they meet all the demands of the modern system [4].

A 100-GB digital coherent network is already deployed in many countries, including Japan, while research work for 400 GB is still ongoing for commercial implementation [4]. As DSP plays a vital role in better signal integrity in high-speed systems, research into equalization techniques is also popular for future systems [6, 7, 29, 30].

4.3.1 Proposed System Design and Analysis

A digital coherent RoF system can be divided into three main parts: (i) a coherent transmitter, (ii) a coherent receiver, and (iii) a RoF unit. Figure 4.10 presents a block diagram of the proposed coherent RoF system, which is based on the same opt-electrical components as other typical opt-electrical devices. A LinBo3 or an MZ modulator is used for the upconversion of electrical signals to optical signals. The modulator is the main component on the transmitter side, and many previously discussed techniques work on it. Ninety-degree hybrids and LO are front-end components for the coherent receiver. The information from

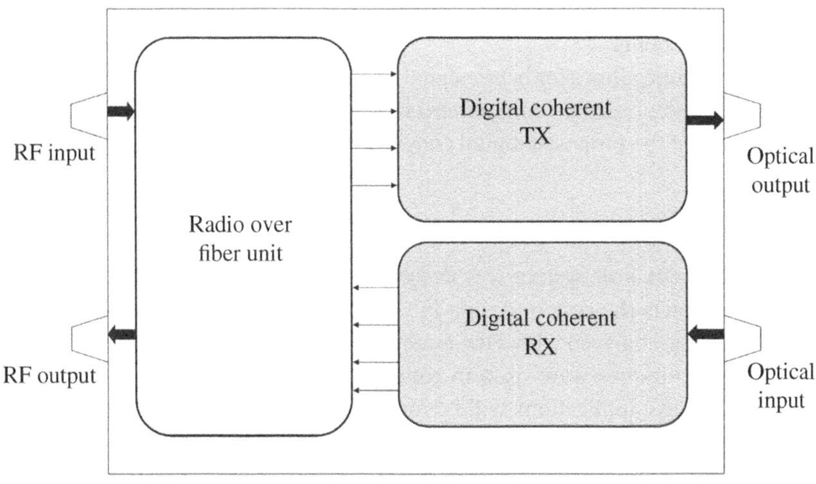

Figure 4.10 Proposed digital coherent RoF block diagram.

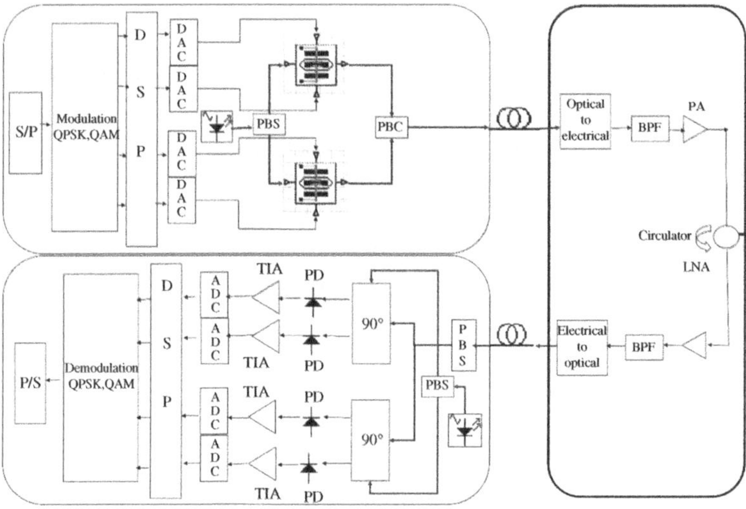

Figure 4.11 Proposed digital coherent radio over fiber system level architecture.

the received light can be extracted after mixing it with LO light signals. Then, a photo detector will receive the information in the form of light and transfer it to a trans-impedance amplifier after down conversion to an electrical signal. For better signal integrity, the ADS passes the received data to a DSP module. We can consequently say that coherent transmitters have passive optical components such as polarized beam splitters and silica light-based circuits, while receivers have high-speed photo detectors, TIA, and ADS, all of which are high-speed electrical components.

The DSP module allows for better signal integrity. Optical-electrical converters, PAs, and LNAs are the main components of a RoF unit. Figure 4.11 illustrates the system design of the proposed digital coherent RoF system.

4.3.2 Simulation

A digital coherent RoF system was designed and simulated in OptiSystem 15. Figure 4.12 depicts the system design in OptiSystem for simulation. The system performance was analyzed with a data rate of 110 GB, and DP-QPSK and 16-QAM modulation techniques were used to compare the performance for 60–140 km. As indoor wireless applications will be multifrequency applications, system performances were also analyzed with 2.4 and 5 GHz frequencies. The performance

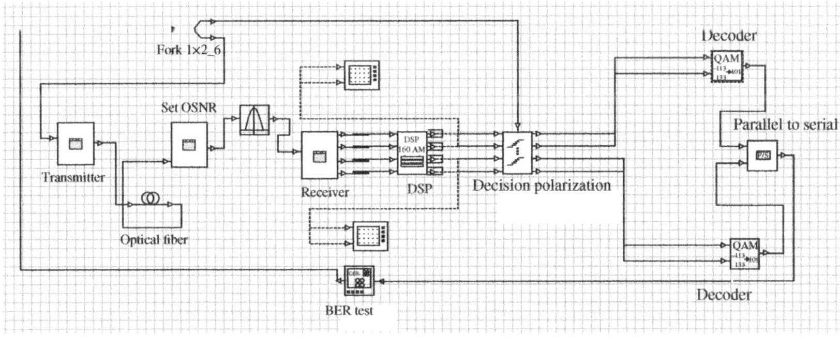

Figure 4.12 Digital coherent radio over fiber system design in OptiSystem.

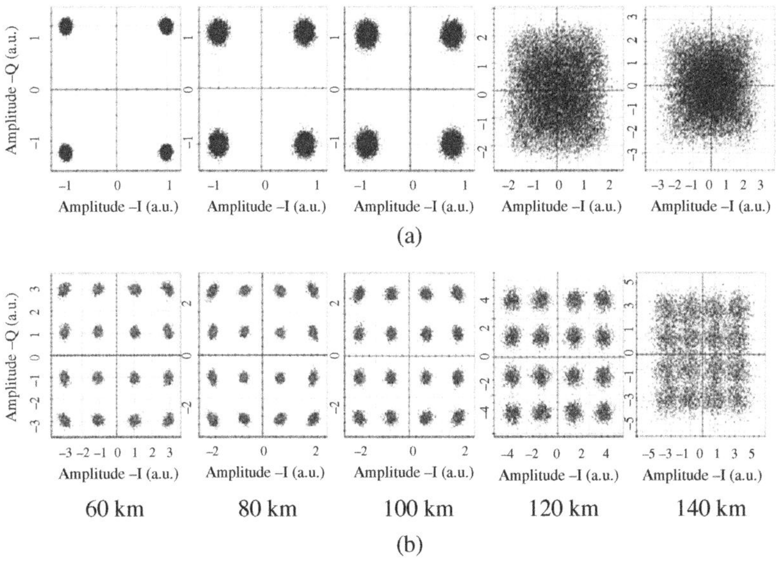

Figure 4.13 110 GB, 60–140 km (a) DP-QPSK (b) 16 QAM.

analyses of both modulation techniques are presented in Figure 4.13. The 16-QAM modulation demonstrated better performance throughout the simulation from 60 to 140 km. The output eye diagram for 2.4 and 5 GHz is presented in Figure 4.14. From the results, we can say that digital coherent RoF can be integrated into future indoor high-speed wireless and wire-line systems to support high-speed backhaul connectivity.

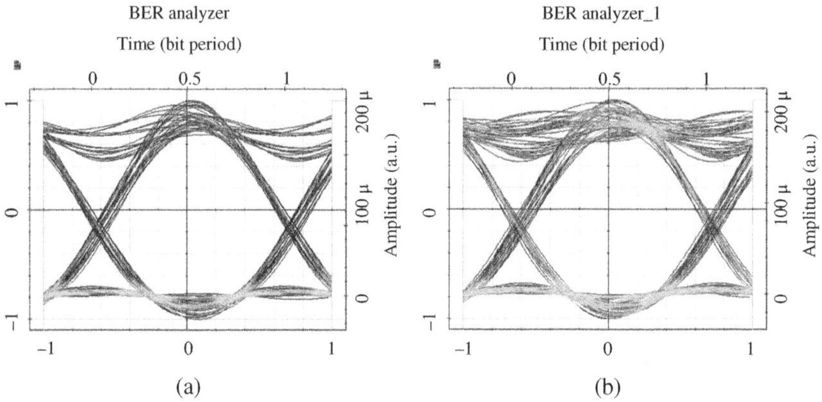

Figure 4.14 Eye diagram (a) 2.4 GHz (b) 5 GHz.

References

1 Y. Miyamoto, "Over 400 Gbit/s digital coherent channels for optical transport network," in *OptoElectronics and Communications Conference (OECC) held jointly with 2016 International Conference on Photonics in Switching (PS)*, pp. 1–3, 2016.

2 M. Poland, C. Nugent, H. Wang, and L. Chen, "Smart home research," *Int. J. Ambient Comput. Intell.*, vol. 1, no. 4, pp. 32–45, 2009.

3 P. Cochrane, "Fiber-to-the-home (FTTH) costs are now in!," *Proceedings of the IEEE*, vol. 96, no. 2, pp. 195–197, 2008.

4 N. J. Frigo, P. P. Iannone, and K. C. Reichmann, "A view of fiber to the home economics," *IEEE Commun. Mag.*, vol. 42, no. 8, 2004, 16–23.

5 S. J. Savory, G. Gavioli, R. I. Killey, and P. Bayvel, "Electronic compensation of chromatic dispersion using a digital coherent receiver," *Opt. Express*, vol. 15, no. 5, pp. 2120–2126, 2007.

6 S. Savory, "Digital coherent optical receivers: algorithms and subsystems," *IEEE J. Sel. Top. Quantum Electron.*, vol. 16, no. 5, pp. 1164–1179, 2010.

7 Y. Mori, C. Zhang, M. Usui, K. Igarashi, K. Katoh, and K. Kikuchi, "200-km transmission of 100-Gbit/s 32-QAM dual-polarization signals using a digital coherent receiver," *Electr. Eng.*, vol. 17, no. 3, pp. 3–4, 2009.

8 S. Chandrasekhar and X. Liu, "Experimental investigation on the performance of closely spaced multi-carrier PDM-QPSK with digital coherent detection," *Opt. Express*, vol. 17, no. 24, pp. 21350–21361, 2009.

9 S. J. Savory, "Digital filters for coherent optical receivers," *Opt. Express*, vol. 16, no. 2, p. 804, 2008.

10 S. J. Savory, "Compensation of fibre impairments in digital coherent systems," in *European Conference on Optical Communication, ECOC*, 2008.
11 X. Liu, S. Chandrasekhar, and A. Leven, "Digital self-coherent detection," *Opt. Express*, vol. 16, no. 2, p. 792, 2008.
12 D. Lavery and S. J. Savory, "Digital coherent technology for long-reach optical access," in *Conference on Optical Fiber Communication, Technical Digest Series*, 2014.
13 K. Kikuchi, "Polarization-demultiplexing algorithm in the digital coherent receiver," *2008 Digest of the IEEE/LEOS Summer Topical Meetings*, no. 9, pp. 101–102, 2008.
14 K. Kikuchi, "Digital coherent optical communication systems: Fundamentals and future prospects," *IEICE Electron. Express*, vol. 8, no. 20. pp. 1642–1662, 2011.
15 B. Spinnler, "Complexity of algorithms for digital coherent receivers," *2009 35th European Conference on Optical Communication*, no. 1, pp. 3–6, 2009.
16 K. Roberts, D. Beckett, D. Boertjes, J. Berthold, and C. Laperle, "100G and beyond with digital coherent signal processing," *IEEE Commun. Mag.*, vol. 48, no. 7, pp. 62–69, 2010.
17 G. Li, "Recent advances in coherent optical communication," *Adv. Opt. Photonics*, vol. 1, no. 2, p. 279, 2009.
18 R. Andres Soriano, F. N. Hauske, N. Guerrero Gonzalez, Z. Zhang, Y. Ye, and I. Tafur Monroy, "Chromatic dispersion estimation in digital coherent receivers," *J. Light. Technol.*, vol. 29, no. 11, pp. 1627–1637, 2011.
19 K. Kikuchi and K. Igarashi, "Characterization of semiconductor-laser phase noise with digital coherent receivers," in *Optical Fiber Communication Conference and Exposition (OFC)*, pp. 1–3, 2011.
20 C. Laperle and M. Osullivan, "Advances in high-speed DACs, ADCs, and DSP for optical coherent transceivers," *J. Light. Technol.*, vol. 32, no. 4, pp. 629–643, 2014.
21 R. Maher and B. Thomsen, "Dynamic linewidth measurement technique using digital intradyne coherent receivers," *Opt. Express*, vol. 19, no. 26, p. B313, 2011.
22 K. Kikuchi and S. Tsukamoto, "Evaluation of sensitivity of the digital coherent receiver," *J. Light. Technol.*, vol. 26, no. 13, pp. 1817–1822, 2008.
23 C. Bellanger, A. Brignon, J. Colineau, and J. P. Huignard, "Coherent fiber combining by digital holography," in *CLEO/Europe – EQEC 2009 – European Conference on Lasers and Electro-Optics and the European Quantum Electronics Conference*, 2009.
24 K. Kikuchi, "Coherent detection of phase-shift keying signals using digital carrier-phase estimation," *2006 Optical Fiber Communication Conference and the National Fiber Optic Engineers Conference*, vol. 24, no. 1, p. OTuI4, 2006.

25 F. Derr, "Coherent optical QPSK intradyne system: Concept and digital receiver realization," *J. Light. Technol.*, vol. 10, no. 9, pp. 1290–1296, 1992.
26 L. E. Nelson and S. L. Woodward, "Capabilities of real-time digital coherent transceivers," in *European Conference on Optical Communication, ECOC*, vol. 1–2, 2010.
27 R. Borkowski, D. Zibar, and I. Tafur Monroy, "Anatomy of a digital coherent receiver," *IEICE Trans. Commun.*, vol. E97–B, no. 8, pp. 1528–1536, 2014.
28 T. Pfau, S. Hoffmann, and R. Noé, "Hardware-efficient coherent digital receiver concept with feedforward carrier recovery for M-QAM constellations," *J. Light. Technol.*, vol. 27, no. 8, pp. 989–999, 2009.
29 S. Savory, "Digital signal processing for coherent systems – OSA technical digest," *Optical Fiber Communications Conference*, p. OTh3C.7, 2012.
30 T. S. Koichi Murata, "Optical device technologies for future network evolution, NTT technical review," *NTT Photonics Laboratories*, p. Japan, 2011.
31 Zhou, X. and Yu, J., "Digital signal processing for coherent optical communication," in *2009 18th Annual wireless and optical communications conference*, 2009, pp. 1–5.
32 M. Hanna, "Is DP-QPSK the end game for 100 Gbits/sec?," Light wave, 2008. [Online]. Available: http://www.lightwaveonline.com.
33 R. Kashif, O. J. Famoriji, and F. Lin, "Equalization techniques to ensure signal integrity in high speed serial and optical design," in *2016 IEEE International Conference on Ubiquitous Wireless Broadband*, ICUWB 2016, 2016, pp. 3–5.

5

Proposed Hybrid System for Indoor VLC

5.1 Overview

Indoor wireless communication is rapidly rising, and an increasing number of devices inside our homes are connected to the Internet. The Internet of Things (IoT) will open new doors in connectivity. Smart homes, offices, and campuses all consume a vast range of the radio spectrum and cause spectrum crunch. The Ethernet currently deployed inside buildings and homes will no longer be powerful enough to maintain the data speed of future technologies such as fifth generation (5G) and even sixth generation (6G), which will be fiber and satellite high-speed communication. From indoor to outdoor environments, there will be a need for high-speed, high-capacity, and secure communication. Optical fiber can provide a wide range of data and security to future networks. A prospective future scenario is depicted in Figure 5.1. Visible light communication (VLC) can be integrated into other wireless technologies such as radio frequency (RF), IR, Bluetooth, and WiMax. In the previous section, we mentioned that VLC has some limitations in different situations. To overcome these limitations, we can combine a few currently existing technologies to create a hybrid system and to enjoy the benefits of existing technologies along with VLC. To date, most of the research work centers around designing techniques for hybrid networks and involves channel modeling; however, there is a lack of system architecture. Therefore, we need a system design for the hybrid network, and that system should be capable of an indoor connection network with high-speed backhaul while also providing uplink connectivity for Li-Fi. Figure 5.2 provides an overview of a hybridized system in an indoor scenario. Backhaul connection and uplink connectivity are the two goals for the design of a hybridized system.

Hybrid Communication Systems for Future 6G and Beyond: Visible Light Communication & Radio over Fiber Technology, First Edition. Rao Kashif.

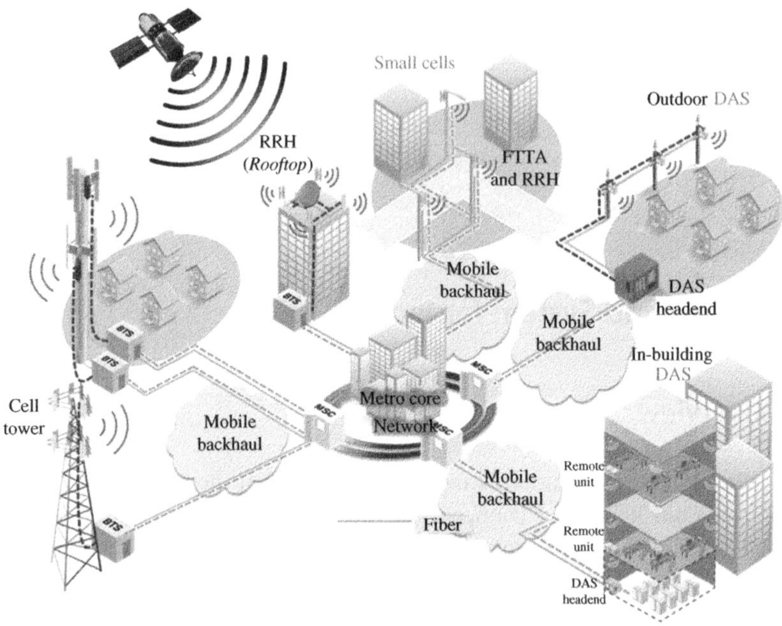

Figure 5.1 Future hybrid fiber network.

5.1.1 Backhaul Connection

The OFDM coherent RoF-based hybridized Wi-Fi/Li-Fi system architecture is proposed. Future Li-Fi networks require a backhaul system that can hold high-speed communication, which is not possible with traditional copper-wired systems. Therefore, the proposed architecture of the hybridized system can be helpful for future FTTH networks, and it will be a useful candidate for holding the high Li-Fi data rate.

5.1.2 Uplink Connectivity

In Li-Fi systems, IR is currently available inside the modules of Li-Fi units. Numerous IR uplink approaches are proposed [1, 2]. IR can be an ideal option in RF-restricted areas; however, it is not suitable for dense indoor environments because people are not in stationary positions, and Wi-Fi can thus be a more suitable option for providing uplink services in the hybridized system. In this dissertation, an architecture of the auto channel switching unit (ACSU) is proposed. This unit will be responsible for channel switching and providing uploading through a Wi-Fi channel and downloading through Li-Fi. Downloading traffic can also be a switch to Wi-Fi in cases of Li-Fi unavailability.

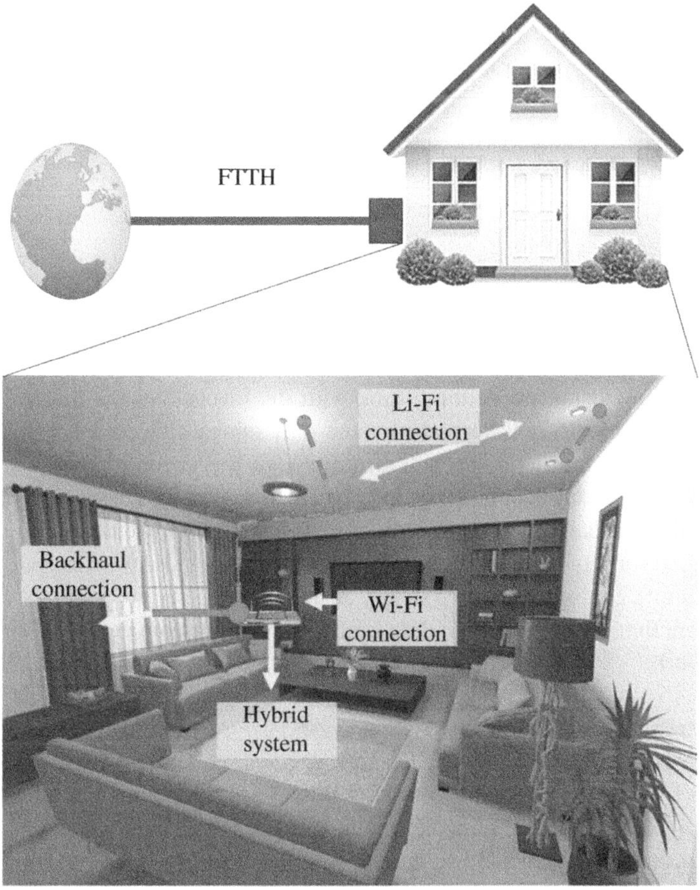

Figure 5.2 Indoor visible light hybrid system.

5.2 Proposed System Design

Figure 5.3 presents an overview of the proposed hybridized system. This system is based on three main blocks: (i) an OFDM coherent RoF TX, (ii) an OFDM coherent RoF RX, and (iii) ACSU.

5.2.1 OFDM Coherent RoF

Copper can deliver a high bandwidth only over a short distance. On the one hand, we can obtain high bandwidths through copper only if the service provider is near our home. However, on the other hand, optical fiber is famous for its high

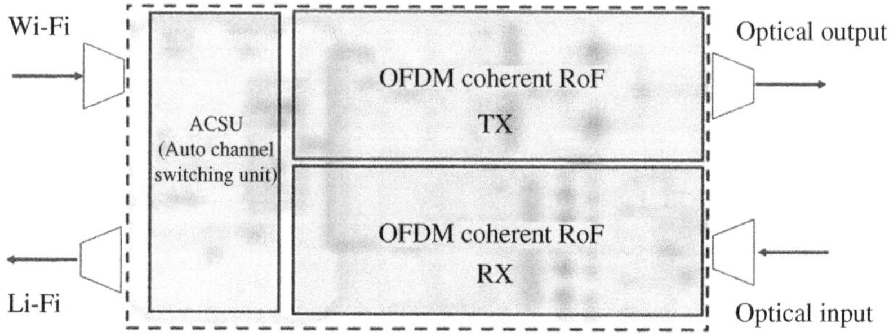

Figure 5.3 Proposed hybrid system block diagram.

bandwidth over long distances. In the future, with the involvement of the IoT, the demand for bandwidth will keep rising, and only optical fiber can provide enough bandwidth to fulfill the need for future high-speed communication. Some of the high-bandwidth applications are HD video streaming/downloading, telemedicine, distance learning, and telecommuting. One HD movie takes up to 35,000 web pages worth of bandwidth. While VLC is the best candidate to support such high data-rate communication systems, an indoor VLC system requires a backhaul system that can handle high data traffic and provide connectivity with an FTTH network.

5.2.1.1 Architecture Design

The proposed system can be categorized into six parts: coherent and OFDM transceivers, a DSP, and visible light front-end units. The system provides backhaul connectivity with an FTTH network on one side, and another side has VLC front-end connectivity, as depicted in Figure 5.4. The transmitter sends a signal to the digital coherent receiver via the optical channel, and after the signal is received in light form, it is mixed with the local light oscillator. Photo detectors

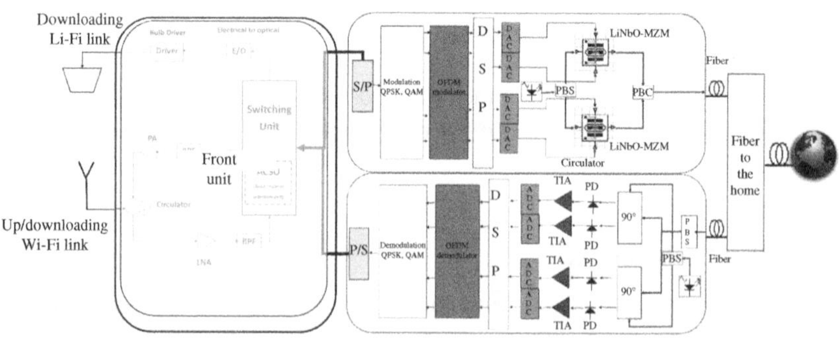

Figure 5.4 Proposed hybrid system architecture.

then pass over information to an ADC, which will be further processed by a DSP and decision portion The transmitter has both optical and electrical components. Furthermore, the modulator is an essential component of a transceiver, as it combines electrical information and light signals for an optical channel.

The front part of the coherent receiver has 90° hybrids and LO. The information is extracted by mixing the received light with LO light – LO provides signal gain and its shot noise sweeps over receiver thermal noise. Thereafter, a photo detector receives data from light and forwards information to transimpedance amplifiers and further to ADS, which will hand over digital information to a DSP chip for signal integrity. By knowing this process, we can see that a digital coherent optical receiver is based on three parts, other than a DSP chip. The first portion is based on passive optical circuits having polarized beam splitters and silica light-based circuits.

The second part consists of high-speed optical semiconductors, which are ultra-high-speed photo detectors, and the last one is made up of trans-impedance amplifiers, which are high-speed electronic devices. In the end, an OFDM signal will be demodulated.

5.2.2 Modeling in OptiSystem 15

Radio over fiber (RoF) is the best candidate to provide high-speed multimedia signals for high-definition video and other wireless communication applications [3]. The modulation format is the critical element for multi-gigabit optical systems, and to reach such a capacity, an OFDM can be used because of its multicarrier modulation format [4]. It can also reduce chromatic dispersion in optical fiber [5]. An OFDM coherent RoF has five blocks: (i) RF-OFDM Tx, (ii) RF-to-optical up convertor, (iii) optical link, (iv) optical-to-RF down converter, and (v) RF-OFDM RX. Table 5.1 lists the parameters for the OFDM coherent RoF simulation.

The whole system was simulated in the OptiSystem 15.0 commercial simulation tool. Figure 5.5 illustrates the simulated system layout. At the transmitter side, the LiNbO-MZM modulator is used inside a digital coherent TX, as indicated in Figure 5.6. It is an essential component between RF and laser for a high bit rate [6]. The output field formula for LiNbO-MZM is as follows [7]:

$$E_0(t) = \frac{E_{(\text{in})}(t)}{10^{\left(\frac{\text{IL}}{20}\right)}}[\gamma \cdot e^{(\text{Em})} + (1-\gamma) \cdot e^{(\text{Em})}] \tag{5.1}$$

$$\text{Em} = {}^{j\pi v_2(t)}/_{V_{(\pi\text{RF})}} + {}^{j\pi v_{(\text{bias }2)}}/_{V_{(\pi\text{DC})}} \tag{5.2}$$

where $E_{\text{in}}\,(t)$ is the optical input signal, IL is the parameter inserting loss, $v_1\,(t)$ and $v_2\,(t)$ are the input electrical voltages, $V_{\text{bias }1(t)}$ and $V_{\text{bias }2(t)}$ are set for bias voltage 1 and 2, $V_{\pi\text{RF}}$ and $V_{\pi\text{DC}}$ are the switching modulation and bias voltage respectively, and γ is the power splitting ratio of both Y-branch waveguides.

Table 5.1 Simulation parameters.

Modulation	DP-QPSK, 4-QAM, 16-QAM
Bit rate	10 Gb/s
CW laser freq.	193.1 THz
Fiber length	30, 50, 70 km
Attenuation	0.2 dB/km
LO power	−2 dB/m
Photodiode responsiveness responsively	1 A/W

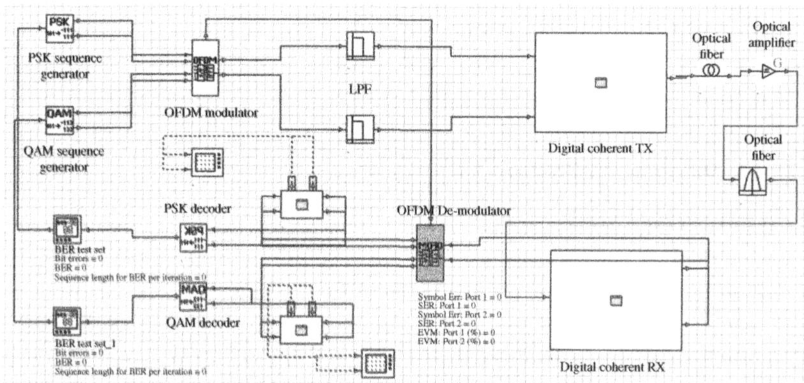

Figure 5.5 OFDM coherent RoF system modeling in OptiSystem.

On the receiver side, the first block is an optical-to-RF down converter, as illustrated in Figure 5.7. At the front-end, the photodiodes are used to convert the optical signal into an electrical current. Here, we conducted performance analyses of APD and PIN photodiodes, depicted in Figure 5.8, and Figure 5.9 presents the Q-factor analyses of both photodiodes. The APD is more suitable for long-haul communication due to high SNR. References [2, 3, 8–10] also experimentally proved the better performance of APD. In short, PIN is more suitable for short-distance applications. Moreover, the OFDM coherent RoF receiver is just the inverse of the transmitter. System simulations were done with three modulation techniques: (i) 4-QAM (ii) 16-QAM, and (iii) QPSK for 30, 50, and 70 km. The output is presented in signal constellation form in Figure 5.10. The BER analysis demonstrated that 16-QAM has a lower BER value, compared to 4-QAM and QPSK.

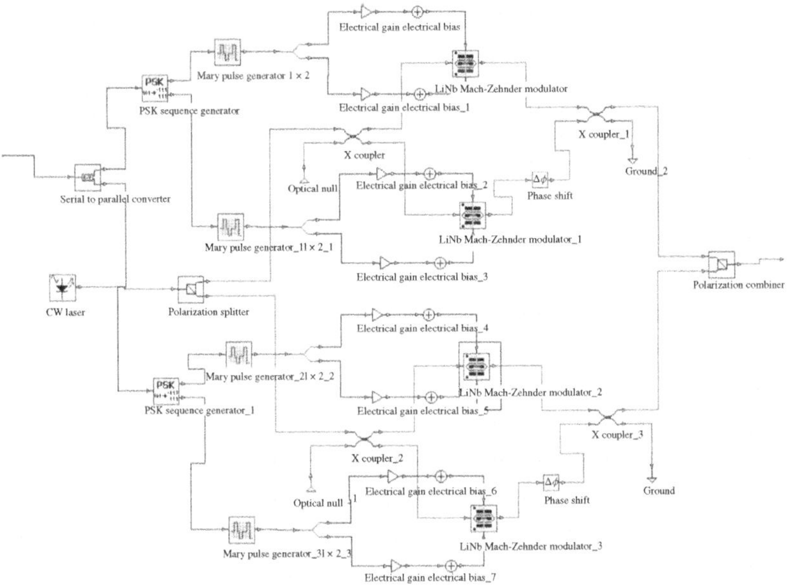

Figure 5.6 Coherent TX.

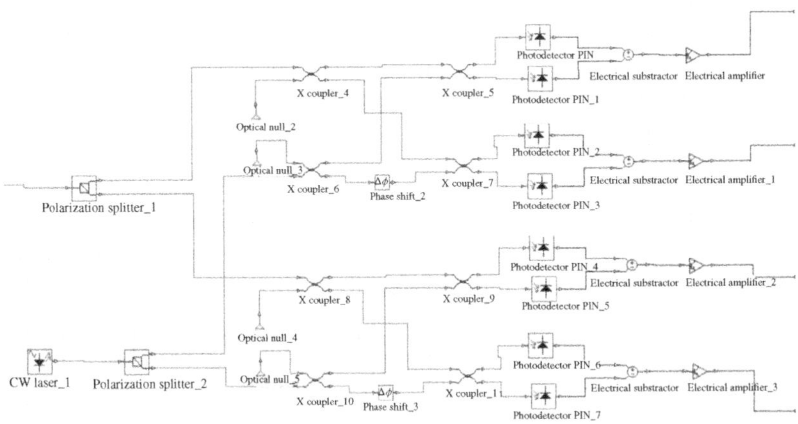

Figure 5.7 Coherent RX.

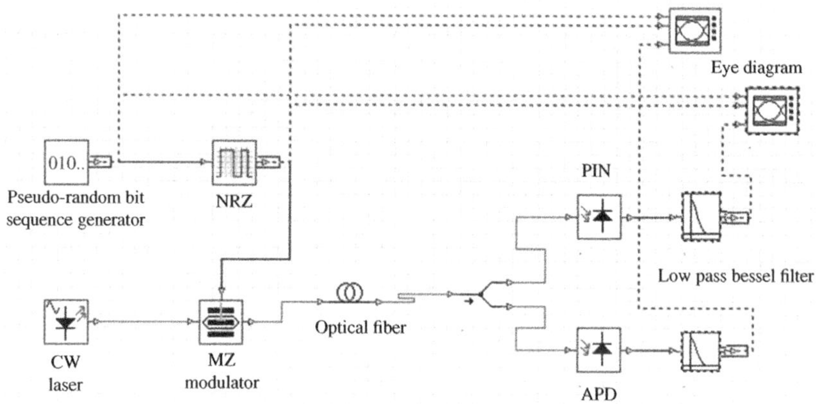

Figure 5.8 APD and PIN diode performance analysis.

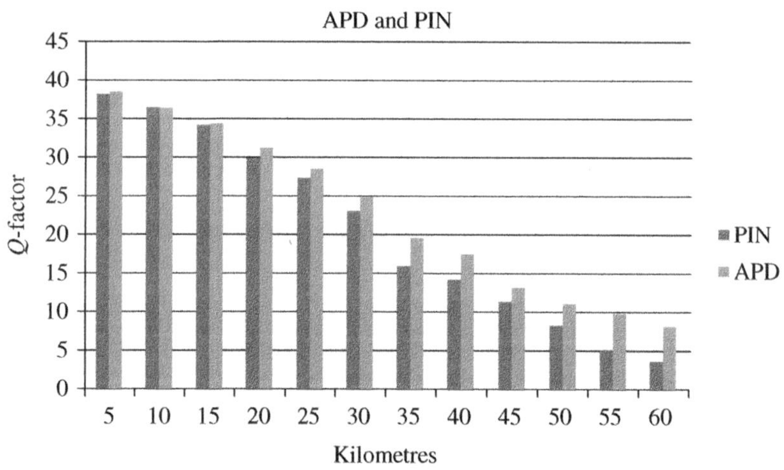

Figure 5.9 PIN and APD *Q*-factor analysis.

The backhaul connectivity issue for indoor VLC is highlighted, and an OFDM-based coherent RoF system architecture design is proposed since it can provide high-speed connectivity to an indoor VLC system with an outdoor FTTH network. The system was also simulated in OptiSystem 15, which is one of the best commercial tools for optical system design. The proposed system could be a prime candidate for providing high-speed backhaul connectivity for indoor VLC systems.

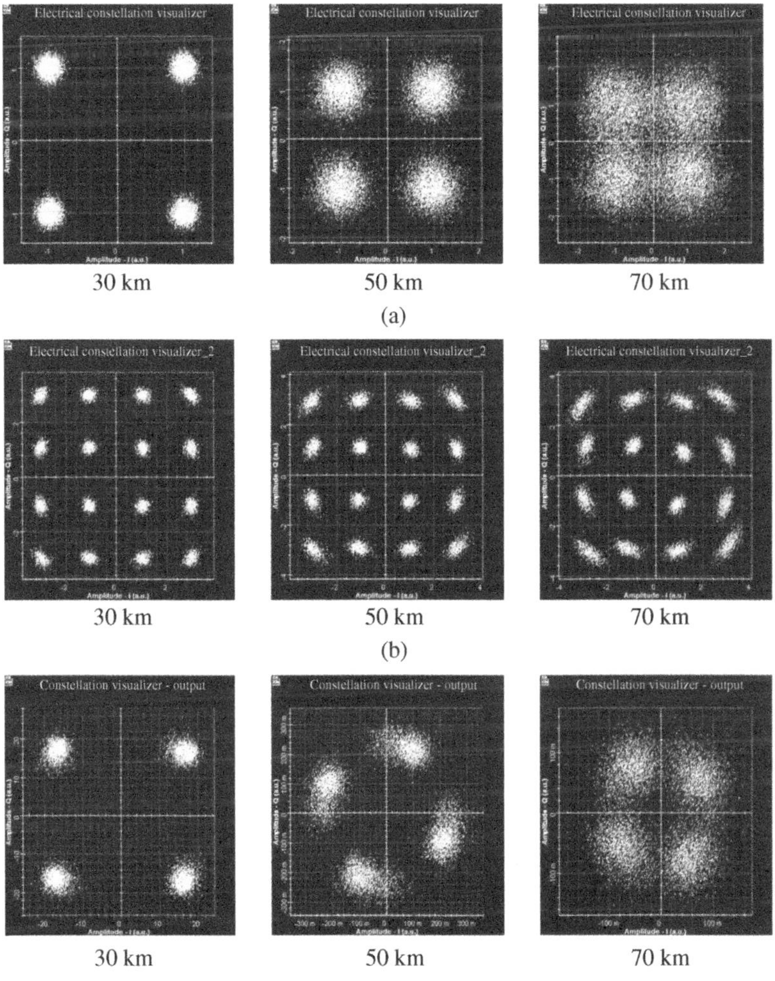

Figure 5.10 Signal constellation results: (a) DP-QPSK, (b) 16-QAM, (c) 4-QAM.

5.3 Proposed Auto Channel Switching Unit (ACSU)

The ACSU is the main block in this hybrid system. It is responsible for routing traffic on the desired channel. The block diagram of a hybrid system with ACSU is presented in Figure 5.11. The typical function of this unit is to handle Li-Fi downloading traffic and Wi-Fi uploading traffic. Moreover, the ACSU also handles the provision of full duplex Wi-Fi communication when Li-Fi is not available.

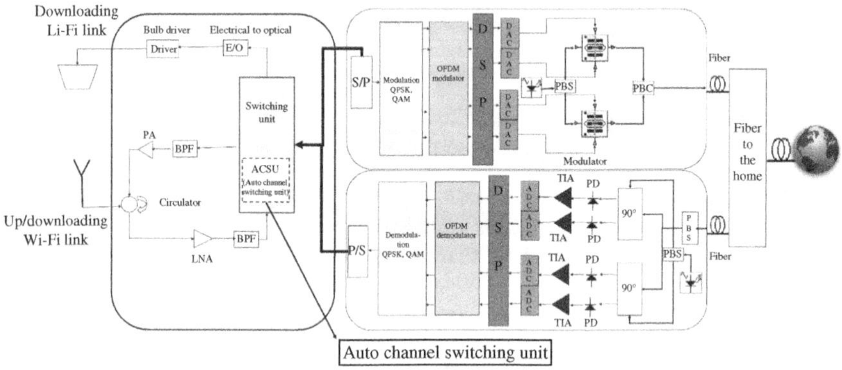

Figure 5.11 Hybrid system with auto channel switching unit.

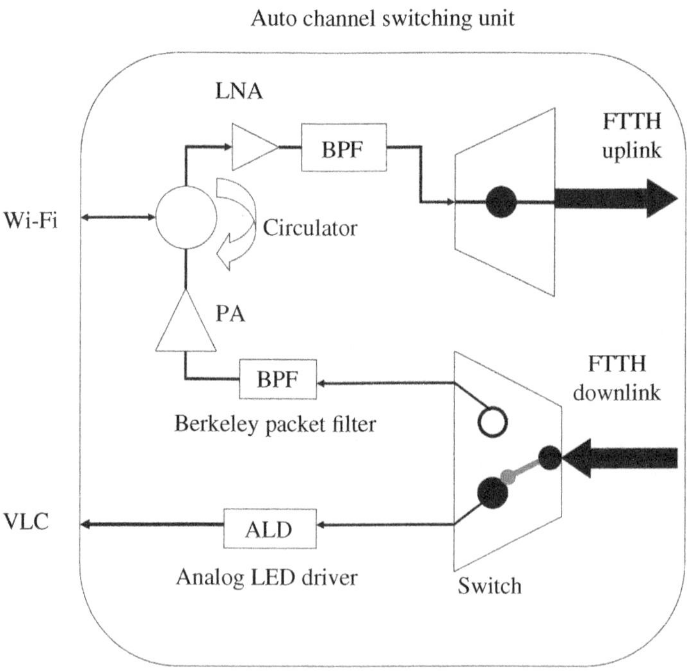

Figure 5.12 Auto channel switching unit architecture.

Access point selection in a heterogeneous network is more complicated, compared to homogeneous networks [11]. An overview of the architecture of the ACSU is presented in Figure 5.12. There is much ongoing research on load sharing in a heterogeneous network, and few works employ the same methods as those of a homogeneous network [12, 13]. Wu et al. [11] proposed a novel APS method for

Table 5.2 Bandwidth for different applications.

No#	Application	Mbps
1	E-mail	0.4
2	Web browsing	0.4
3	HD video streaming	4
4	Audio streaming	2
5	Online gaming	4
6	Video call	0.3
7	Audio call	0.03

a hybrid Li-Fi and Wi-Fi network by using fuzzy logic, and an ACSU is a suitable platform for the practical use of such a technique. As in load balancing field of the area still needs time for getting mature, so we started an ACSU unit with simple rules for the sharing and switching of traffic. Figure 5.14 provides an overview of the ACSU process. In the proposed ACSU, the traffic will be routed based on Algorithm 1 and Algorithm 2. Sunlight is a significant issue for VLC, and dark areas where there is either no or little light available also require an alternate; therefore, external light sensor data can be used to switch downloading traffic to a Wi-Fi channel, as indicated in Algorithm 1.

The second common scenario of Wi-Fi failure or conjunction is a lack of bandwidth when the usage numbers increase. An ACSU can distinguish users according to their bandwidth needs. Table 5.2 lists some bandwidth needs in different scenarios. Furthermore, channel traffic can be balanced according to a user's bandwidth need, as in Algorithm 2. Figure 5.13 provides an overview of an

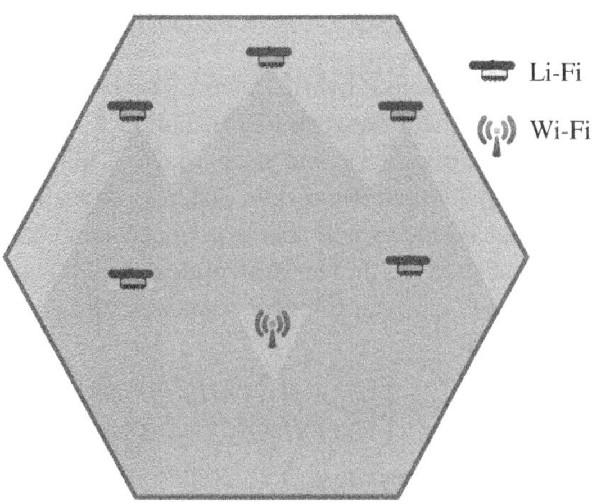

Figure 5.13 Indoor hybrid network.

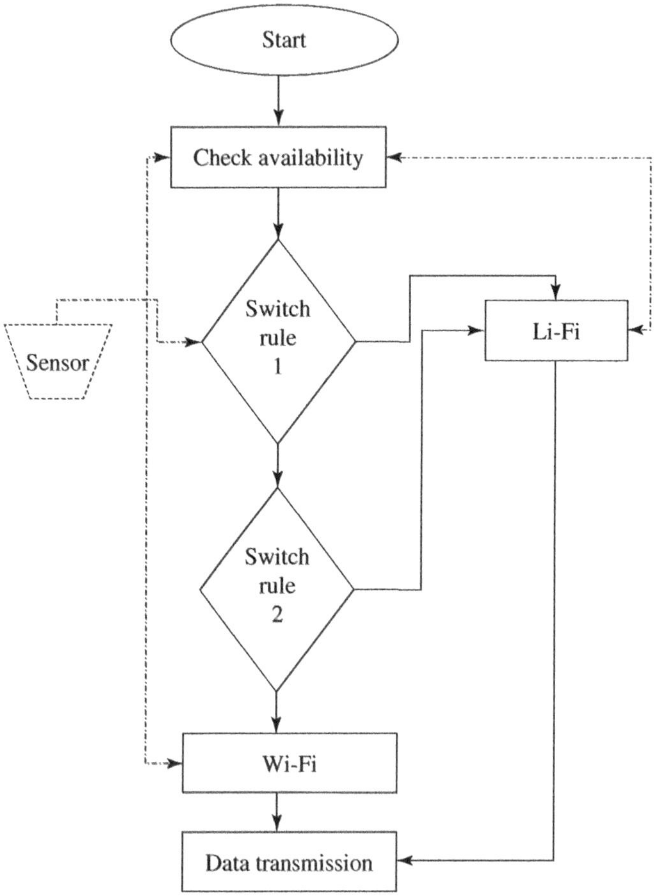

Figure 5.14 Auto channel switching unit flow chart.

ACSU-based indoor hybrid system. The Li-Fi and Wi-Fi APs are connected to an optical backhaul network FTTH. The Li-Fi AP is a large lamp containing LEDs for transmitting data, and users' devices have photodiodes. Each Li-Fi area of coverage is limited to a certain distance, depending on the power of the LED, and APs near windows will also experience some issues in coverage due to outside light and shadows and because optical signals can be blocked.

The Li-Fi channel impulse response can be modulated by the equation from Ref. [14]:

$$H_{\mu a}(f) = \begin{cases} (H_L + H_D(f))H_F(f), & \text{LoS not blocked} \\ H_D(f)H_F(f), & \text{LoS blocked} \end{cases} \tag{5.3}$$

where H_L is the path loss of a LoS channel, $H_D(f)$ is the channel gain by multipath propagation, and $H_F(f)$ is the front-end device frequency response. In Eq. (5.4)

from Ref. [14].

$$H_{\mathrm{L}} = \begin{cases} \frac{(m+1)A_p x^2}{2\pi(z^2+h_{\mathrm{w}}^2)\sin^2\Theta_{\mathrm{F}}} T_{\mathrm{s}}(\theta)\cos^m(\Phi)\cos(\theta), & \theta \leq \Theta_{\mathrm{F}} \\ 0, & \theta > \Theta_{\mathrm{F}} \end{cases} \tag{5.4}$$

m represents the Lambertian index, A_{p} is the photodiode physical area, z is the horizontal distance from the Li-Fi AP, h_{w} is the room height, Φ is the angle of irradiation, θ is the angle of incidence to the photodiode, Θ_{F} is the receiver half angle, T_{s} (θ) is the optical fiber gain, and x represents the refractive index.

According to [15], IEEE 802.11 can have constant maximum throughput for users within 12 m of the AP depicted in Figure 5.14 [16]. Therefore, the system scenario in Figure 5.14 is within the range of 12 m, and Wi-Fi users can obtain maximum throughput.

Algorithm 5.1 Wi-Fi/Light Signal Routing

```
1: function ESTABLISH CONNECTION (A, p, q, r)
2:     Search Device
3:     if Device Free then
4:        if If WiFi Device Free then
5:              Connect WiFi
6:        else if LiFi device free then
7:              Connect LiFi
8:        end if
9:        Send Conformation
10:    else
11:       Check Again Go to Step 3
12:    end if
13: end function
```

Algorithm 5.2 LiFi Routing

```
1: function CHECK LIGHT SENSOR (Light Indoor lx)
2:  if Sensor Available then
3:     if Light Indoor lx < 10 lx or > 77000 lx then
4:        Shift Download Wifi
5:     else if
6:          then stay LiFi
7:     end if
8:     Send Conformation
9:     end if
```

The ACSU will first check the availability of channels, as demonstrated in Algorithm 1 – external sensor data can be used for this process. For a simple scenario, we can assume a maximum of 77,000 LUX (maximum light for reading) and a minimum of 5 LUX (parking area or bedroom low-light scenario).

After passing this stage, the ACSU will make use of Algorithm 2 to check the bandwidth of the channel. In the case of a maximum number of users in both Li-Fi and Wi-Fi APs, the ACSU will distribute users according to the channel data rate. Users can be shifted to another channel according to the data rate requirements. The data rate between the transmitter and user can be expressed by the following equation, as in [16]:

$$R_{\mu,a}^{(n)} = B \log_2 \left(1 + \sin R_{\mu,a}^{(n)}\right) \tag{5.5}$$

The handover time in an indoor scenario is in milliseconds (ms) [17], and it is modeled in the Poisson random process [17]. According to [16, 17], the PMF probability mass function of the handover time is expressed by the following equation from Ref. [17]:

$$P_r(t_{ij} = x) = \frac{\zeta_{ij}^{x} e^{-\zeta_{ij}}}{x!}, \quad x = 0, 1, 2 \ldots \tag{5.6}$$

t_{ij} = overhead switch from AP (i) to AP (j), ζ_{ij} = expectation of overhead.

The handover issue is out of the scope of this work; however, Ref. [17] proposed some handover techniques that can be used in the future in hybrid systems. Algorithms are implemented in the fourth step of Figure 5.14. There are two switching rules proposed. The first switching rule will be based on light sensor data as mentioned in the previous section. If the sensor data is below 5 lx then the data traffic will be transferred to the Wi-Fi channel. While switching rule 2 can be implemented if we use any programmable unit like FPGA or ACSIC device and make the channel limited bandwidth switch for example high-bandwidth applications should remain on the Li-Fi channel and low bandwidth applications are shifted to Wi-Fi. Furthermore, the data traffic will be routed accordingly.

5.3.1 Modeling of the Auto Channel Switching Unit (ACSU)

The main work of the ACSU is to switch the data traffic to either of two channels: Wi-Fi or Li-Fi. To demonstrate this, we can use FPGA, as it is easily programmable for such a purpose. The uploading channel totally relies on Wi-Fi, whereas the downloading channel is controlled by the ACSU.

Therefore, the downloading traffic can be switched by using different channel switching algorithms. A 32-bit signal (sw_in_1) is fed into the demux, as illustrated in Figure 5.15. The signal was switched to one of the channels, namely sw_out_1

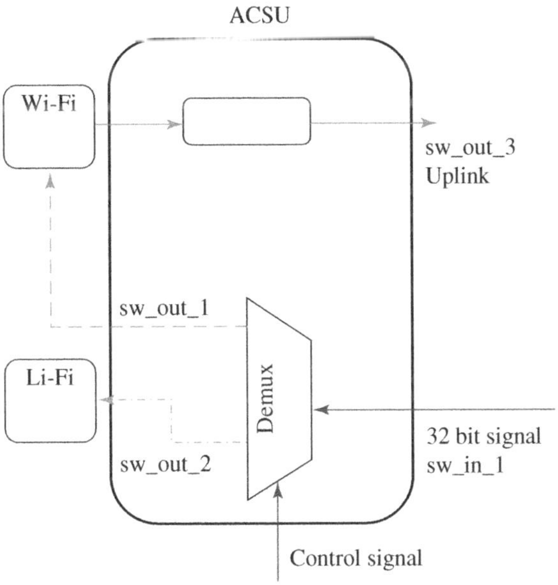

Figure 5.15 ACSU signal flow diagram.

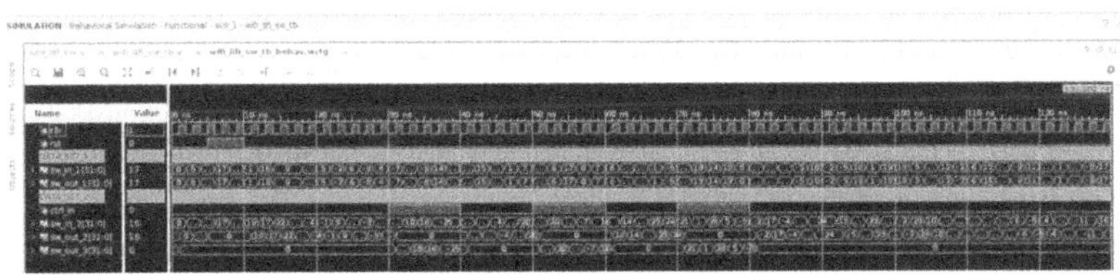

Figure 5.16 ACSU signal simulation.

and sw_out_2, which can be assumed as the Wi-Fi and Li-Fi channels. The signal was switched to one of the two outputs, sw_out_1 or sw_out_2, via algorithms, and in this simulation, we assume the simple scenario that if the Li-Fi channel is off, then the signal will be switched to the Wi-Fi channel. Figures 5.16 and 5.17 display the behavioral signal simulation result. We can see that when the ctrl signal is zero, the data traffic flows through channel sw_out_1, and the data traffic is switched to sw_out_2 when the ctrl signal is 1.

The ACSU plays a vital role in the future hybrid network because by using this unit, we can take advantage of RF and VLC systems. In the future, FPGA can be used as an ACSU, and by connecting with visible light and a Wi-Fi system, a complete demonstration of the hybrid network can be accomplished.

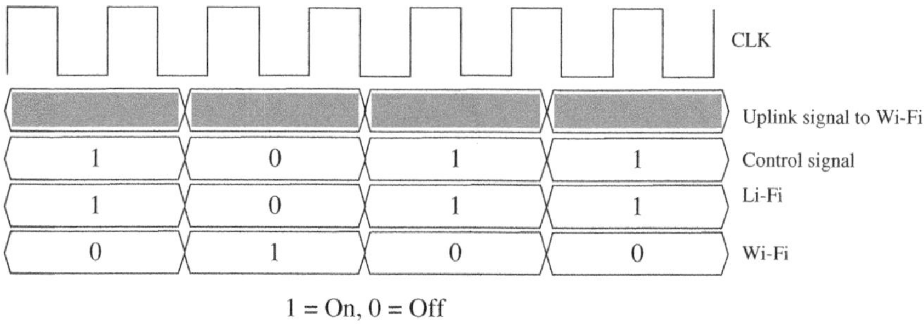

Figure 5.17 Simplifies ACSU signal diagram.

5.4 Feasibility Analysis

Recently almost every industry is introducing hybrid technology due to the number of benefits that can be achieved by utilizing currently existing technologies. The proposed hybridized system also solves backhaul and uplink issues of VLC with the integration of RF technology.

5.4.1 Technical Feasibility

There are some of the factors we need to consider for analysis. (i) Is the solution practical? (ii) Do we have the technology now? So, the answers can be YES. The solution provided for backhaul connectivity is by using OFDM coherent RoF technology and all these three technologies are currently available and also implemented in different areas.

Secondly, the technology proposed for ACSU is also available even though there is a need for integration of some technologies but due to the advancement in FPGA and other ASIC devices, it's possible to bring the concept of ACSU for practical use in future.

5.4.2 Cost-Benefits Analysis

Cost-benefit is one of the advantages we get from hybrid systems. As hybrid systems are the combination of currently existing technologies, the costs of hybrid systems are always low as compared to manufacture or introduce some totally new standalone technology. Indoor optical cables reduce the cost of the whole indoor network by eliminating routers and repeaters from the network which are used in the deployment of Ethernet. Secondly, the uplink hybrid system will use Wi-Fi technology, and currently, every communication device has the ability to access Wi-Fi through built-in Wi-Fi module. Using Wi-Fi for uplink can reduce the cost because users are not going to buy extra infrared plugin devices for uplink through their devices.

References

1 J. Shi, X. Huang, Y. Wang, and N. Chi, "Real-time bi-directional visible light communication system utilizing a phosphor-based LED and RGB LED," in *Wireless Communications and Signal Processing (WCSP), 2014 Sixth International Conference on*, pp. 1–5, 2014.

2 G. Cossu, R. Corsini, A. M. Khalid, and E. Ciaramella, "Bi-directional 400 Mbit/s LED-based optical wireless communication for non-directed line-of-sight transmission," in *Optical Fiber Communications Conference and Exhibition (OFC)*, pp. 1–3, 2014.

3 J. Ma, Y. Zhan, M. Zhou, H. Liang, Y. Shao, and C. Yu, "Full-duplex radio over fiber with a centralized optical source for a 60 GHz millimeter-wave system with a 10 Gb/s 16-QAM downstream signal based on frequency quadrupling," *J. Opt. Commun. Netw.*, vol. 4, no. 7, pp. 557–564, 2012.

4 S. Ahmad and M. Zafrullah, "40 Gb/s 4-QAM OFDM radio over fiber system at 60 GHz employing coherent detection," *J. Mod. Opt.*, vol. 62, no. 4, pp. 296–301, 2015.

5 F. Almasoudi, K. Alatawi, and M. A. Matin, "Performance analysis of 4×10 Gbps OFDM-PON system over ROF link," *Opt. Photonics J.*, vol. 3, no. 2, p. 217, 2013.

6 J. Cartledge, "Performance of 10 Gb/s lightwave systems based on lithium niobate Mach-Zehnder modulators with asymmetric Y-branch waveguides.," *IEEE Photonics Technol. Lett.*, vol. 7, no. 9, pp. 1090–1092, 1995.

7 H. Yang et al., "Evaluation of effects of MZM nonlinearity on QAM and OFDM signals in RoF transmitter," in *Microwave Photonics, 2008. Jointly Held with the 2008 Asia-Pacific Microwave Photonics Conference. mwp/apmp 2008. International Topical Meeting On*, pp. 90–93, 2008.

8 Y. Wang, D. A. Basnayaka, X. Wu, and H. Haas, "Optimisation of load balancing in hybrid LiFi/RF networks," *IEEE Trans. Commun.*, vol. 65, pp. 1708–1720, 2017.

9 P. Gopal, S. Kar, and V. K. Jain, "Performance comparison of PIN and APD based FSO satellite systems for various pulse modulation schemes in atmospheric turbulence," *Commun. Netw.*, vol. 5, no. 3, pp. 200–203, 2013.

10 R. Kashif, F. Lin, and O. John, "Modeling of high speed free space optics system to maintain signal integrity in different weather conditions; system level," *Int. J. Adv. Comput. Sci. Appl.*, vol. 8, no. 3, pp. 45–48, 2017.

11 X. Wu, H. Haas, and M. Safari, "Access point selection for hybrid Li-Fi and Wi-Fi networks," *IEEE Trans. Commun.*, vol. 65, no. 12, pp. 5375–5385, 2017.

12 Y. Wang, H. Haas, X. Wu, "Distributed load balancing for Internet of Things by using Li-Fi and RF hybrid network," in *Personal, Indoor, and Mobile Radio Communications (PIMRC), 2015 IEEE 26th Annual International Symposium on IEEE*, pp. 1289–1294, 2015.

13 X. Li, R. Zhang, and L. Hanzo, "Cooperative load balancing in hybrid visible light communications and Wi-Fi," *IEEE Trans. Commun.*, vol. 63, no. 4, pp. 1319–1329, 2015.

14 L. Wu, Z. Zhang, J. Dang, and H. Liu, "Adaptive modulation schemes for visible light communications," *J. Light. Technol.*, vol. 33, no. 1, pp. 117–125, 2015.

15 M. B. Rahaim, A. M. Vegni, and T. D. Little, "A hybrid radio frequency and broadcast visible light communication system," in *GLOBECOM Workshops (GC Wkshps), 2011 IEEE. IEEE*, pp. 792–796, 2011.

16 Y. Wang and H. Haas, "Dynamic load balancing with handover in hybrid Li-Fi and Wi-Fi networks," *J. Light. Technol.*, vol. 33, no. 22, pp. 4671–4682, 2015.

17 Q. Ye, B. Rong, Y. Chen, M. Al-Shalash, C. Caramanis, and J. G. Andrews, "User association for load balancing in heterogeneous cellular networks," *IEEE Trans. Wirel. Commun*, vol. 12, no. 6, pp. 2706–2716, 2013.

6

Proposed Indoor Hybrid System Modeling

6.1 Modeling of Indoor Hybrid System for VLC

A system-level model was created in OptiSystem 15 and simulated with different scenarios to check the performance of the indoor visible light communication (VLC) system with the hybrid radio over fiber (RoF) system. Figure 6.1 presents the diagram of a hybrid system in OptiSystem. The system can be divided into TX, RX, SMF Channel, Matlab interface, and VLC transceiver. A 16-QAM OFDM signal was generated and sent over a single-mode, 500-m long fiber as an indoor scenario. The signal was demodulated on the receiver, and Matlab was used to retransmit the signal over the visible light channel. The overview of a visible light system in OptiSystem is provided in Figure 6.2. The simulation was repeated with 1-, 2-, and 3-m distances between the VLC TX and RX – 1 m is the normal ceiling height, while 3 m can assume the scenario of high ceiling halls or corridors. Figure 6.3 displays the eye diagrams of all three scenarios, and Figure 6.4 presents the 16-QAM constellation diagram. As we can see, the RoF-based indoor VLC system can provide excellent signal integrity in any indoor scenario. Figure 6.5 illustrates the *Q*-factor analysis, which demonstrated that the *Q*-factor falls with the increase in distance between TX and RX; however, at 3 m, communication is still possible with some power amplification.

6.2 VLC and RoF Indoor Downloading

The experimental scheme for hybrid system downlink transmissions is depicted in Figure 6.6, while Table 6.1 lists the parameters for the experiment. A 52 subcarrier-based 16-QAM signal was generated from vector signal generator (VSG). The data rate was set to 50 Mbit/s, which is enough for high-definition transmission.

Hybrid Communication Systems for Future 6G and Beyond: Visible Light Communication & Radio over Fiber Technology, First Edition. Rao Kashif.

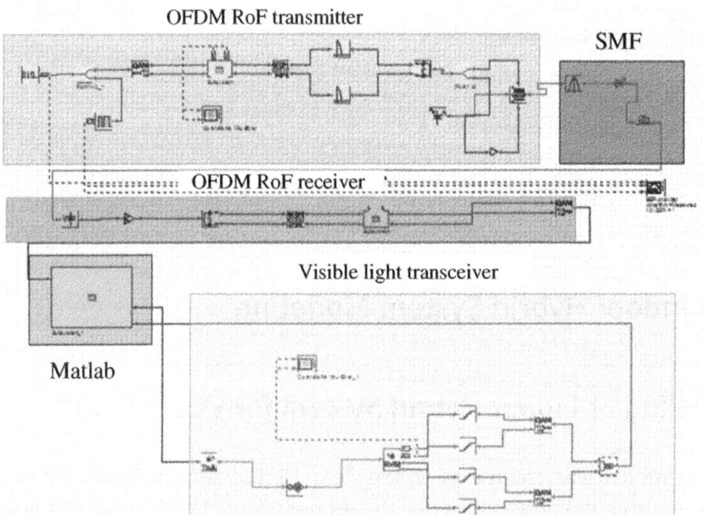

Figure 6.1 Overview of hybrid system.

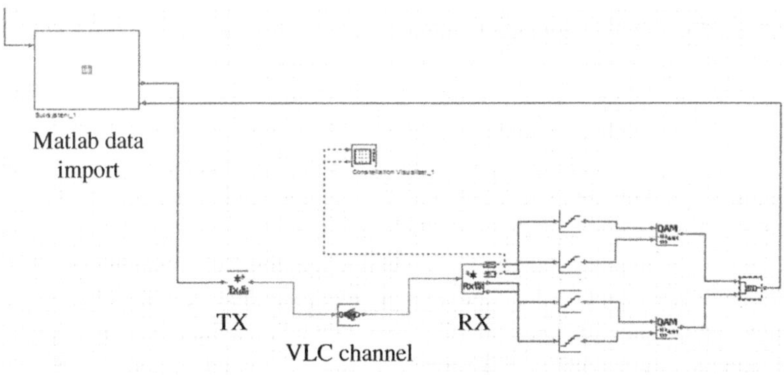

Figure 6.2 VLC system model for hybrid system simulation.

A distributed feedback (DFB) laser of 1550 nm was implemented in the transmitter. Moreover, an SMF of 500 m was used for the medium. Power on the PIN photodiode was controlled by a variable optical attenuator (VOA), and the output electrical signal from the photodiode was down-converted by a mixer and LO at 2.3 GHz. An obtained signal of 25 MHz was amplified and sent to the light emitting diode (LED), which was operated at a 400-mA bias current to achieve the best results. On the receiver side, the TX and RX were placed on proper LoS, and a bi-convex lens and an APD photodiode were used. Furthermore, vector

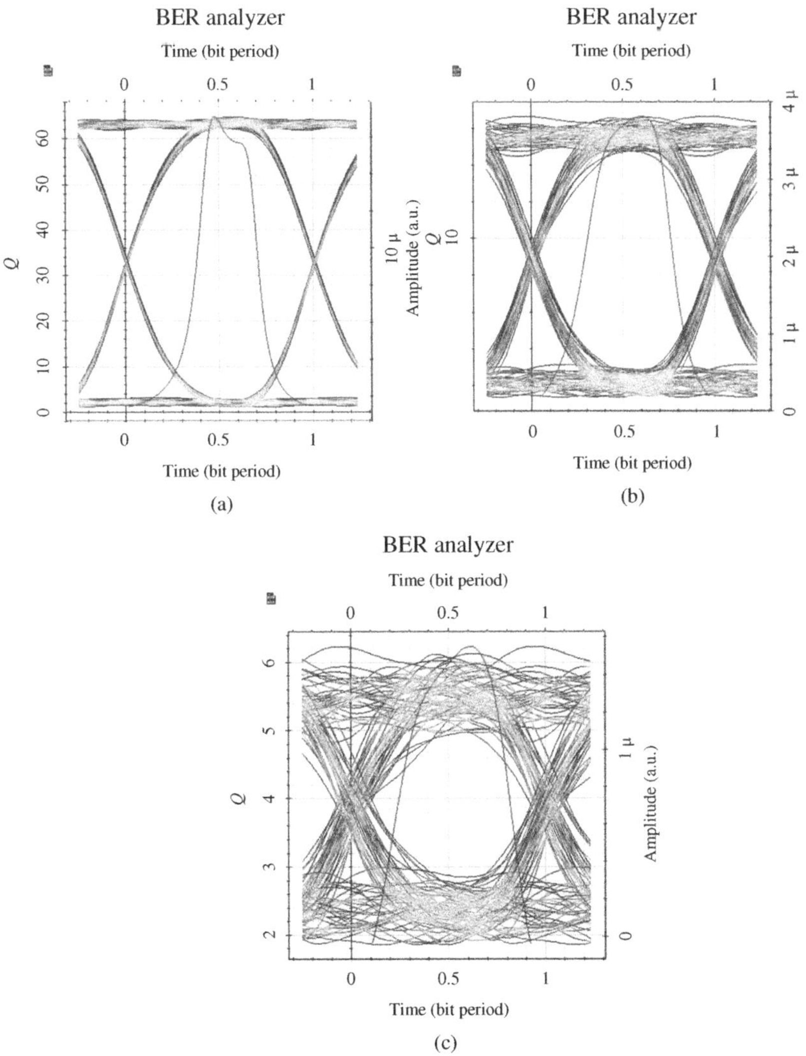

Figure 6.3 Eye diagram (a) 1 m (b) 2 m (c) 3 m.

scale analyzer (VSA) was used to check the quality of the received signal. In the indoor environment, the illumination level is the most important parameter, and it varies based on the distance between the transmitter and receiver. A lx meter was used, along with the APD receiver, to measure the illumination level.

The illumination – or the brightness level – was the most important parameter for our experiment, so we measured the received signal quality in error vector

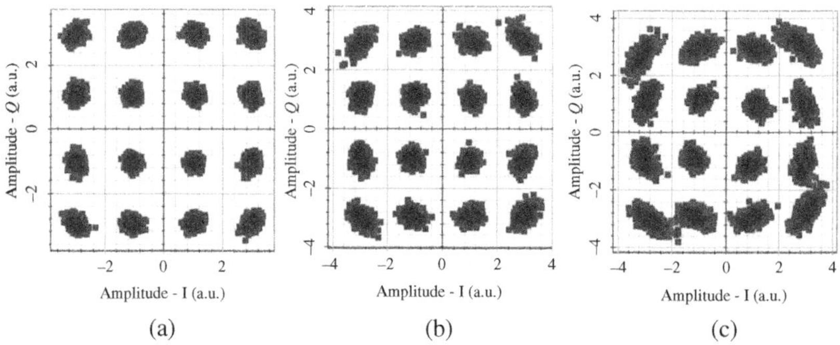

Figure 6.4 16-QAM constellation diagram for (a) 1 m (b) 2 m (c) 3 m.

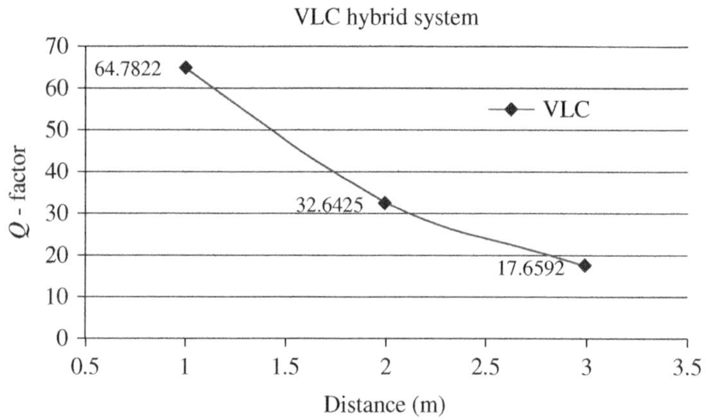

Figure 6.5 Q-factor analysis of VLC.

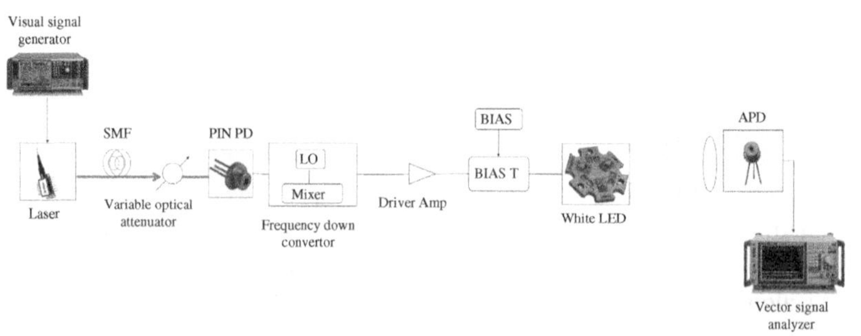

Figure 6.6 Hybrid system downlink transmission.

Table 6.1 Experiment parameters.

Hardware/software	AWG (RFXpress), Oscilloscope (VSA89600)
Fiber	SMF 500 m
Frequency	2.4 GHz
Data rate	50 Mb/s
Modulation	16-QAM
Subcarrier	52, OFDM format

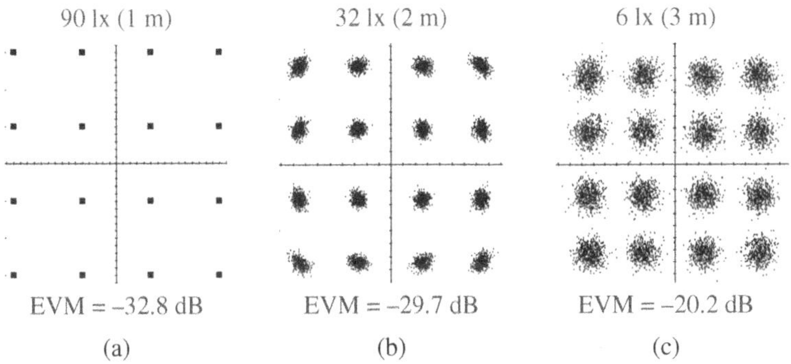

Figure 6.7 16-QAM constellation diagram (a) 1 m (b) 2 m (c) 3 m.

magnitude (EVM) versus the illumination level. The distance between the TX and RX was kept at 1, 2, and 3 m. According to IEEE 8021.11 g for 16-QAM, a signal is EVM −19 dB.

Figure 6.7 displays the 16-QAM constellation diagram for 1, 2, and 3 m. In the case of a distance increase, both the power and the illumination level decrease, which allows one to see the noise in the constellation diagram for 2 and 3 m. Figure 6.8 illustrates the system downlink performance; overall, it was below the limit of IEEE 802.11 g for 16-QAM, which is > −19. It was observed that the system was even working below 10 lx values at 3.5 m.

From the experiment, it is proved that RoF transmitted 50 Mb/s at a distance of 500 m and with a 3.5-m free space distance between the TX and RX, which is the average height of corridors or large conference halls. Also, the performance was under the limit set by IEEE 802.11 g, and it still worked under 10 lx, which can be considered to be low light, for example, a parking area or a bedroom light scenario. Therefore, the experiments demonstrate that RoF-based hybrid systems can be suitable options for future, high-speed, indoor-light broadband and even IoT services.

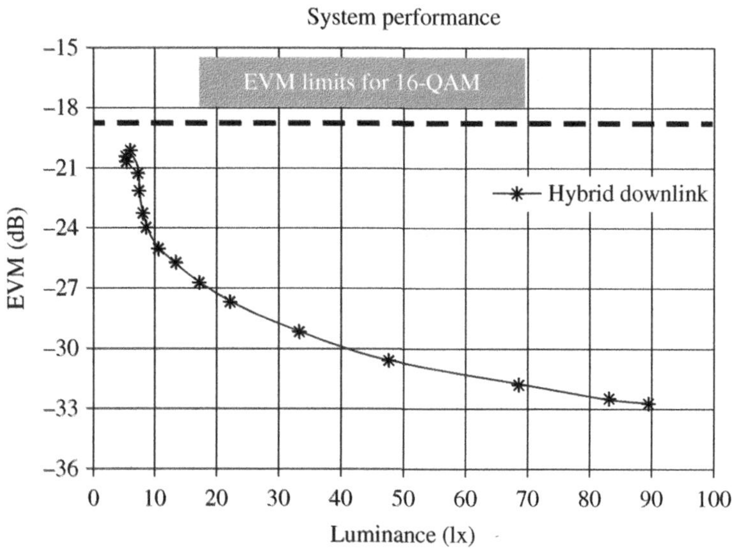

Figure 6.8 Downlink system performance.

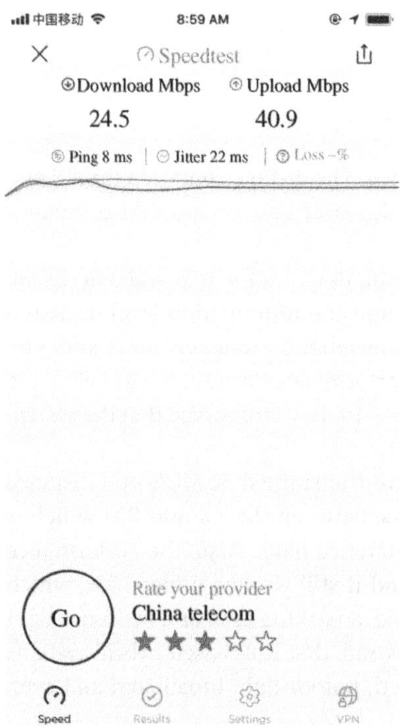

Figure 6.9 Wi-Fi uploading test.

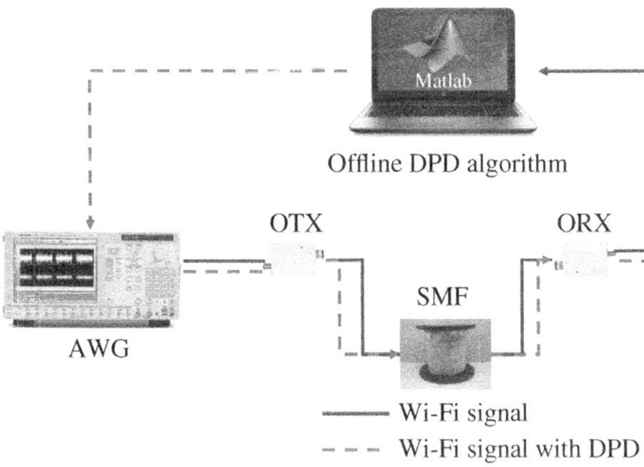

Figure 6.10 Wi-Fi signal over radio over fiber.

6.3 Wi-Fi and RoF for Indoor Purposes

In our proposed system, the uploading in hybrid system will be through Wi-Fi. Few companies currently offer devices with IR; however, due to the availability of IR in devices and also LoS issues, Wi-Fi will be a better option. Also, the current W-Fi uploading speed is sufficient enough to work as an uplink in a hybrid system. Figure 6.9 portrays an uplink test of our lab Internet Wi-Fi connection. To obtain better power in the system, we conducted an experiment for a Wi-Fi signal over indoor RoF.

Figure 6.10 illustrates the experimental setup. A 2.4 GHz, 52-subcarrier, OFDM signal was generated from AWG, and it was transmitted to 500 m through a single-mode fiber cable. The received signal was then analyzed by an oscilloscope.

Coefficients were extracted and processed in the offline Matlab program, and a distorted signal was fed into AWG from the computer. The digital pre-distortion (DPD) was then sent to the same optical medium, and the performance was evaluated by comparing the received EVM of both signals – the EVM can be acquired by VSA89600 software in the oscilloscope. Figure 6.11 presents the performance graph of the Wi-Fi signal with and without DPD.

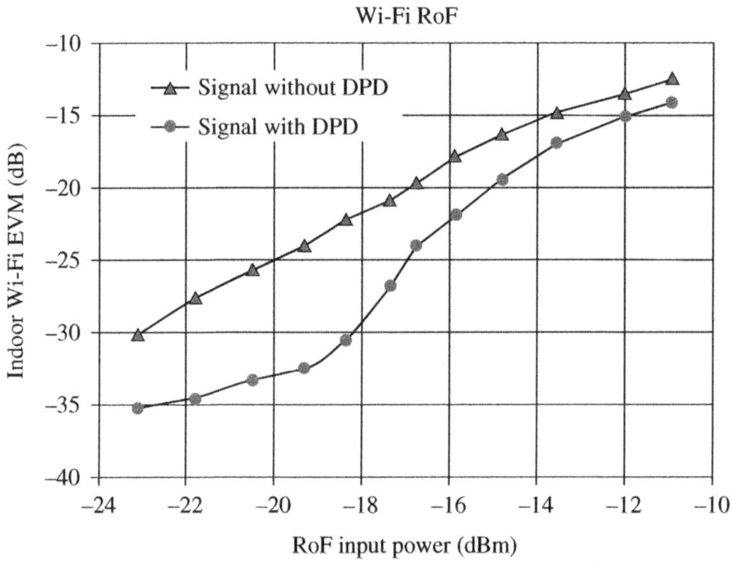

Figure 6.11 Wi-Fi over RoF with and without DPD.

7

Conclusion and Future Work

7.1 Conclusion

This book undertook research regarding the design of a hybrid system for indoor visible light communication (VLC). Visible light techniques currently face a range of challenges, which should be seriously addressed to achieve a wide implementation of visible light. This book focused on indoor hybrid communication systems. A hybrid system, which can provide high-speed backhaul connectivity to an indoor VLC network, was proposed. In addition, an auto channel switching unit (ACSU) design was proposed; it can switch downloading traffic to VLC or a wireless-fidelity (Wi-Fi) channel. The ACSU also provides uplink connectivity to the indoor VLC network via Wi-Fi.

The primary purpose of this book was to carry out in-depth research regarding VLC and to detect technological issues that hybrid communication can solve. To design the hybrid network, a range of systems, including radio over fiber (RoF) and the digital coherent system, were studied. Moreover, the system performance was analyzed. All the designs and analyses regarding the optical system were done through the OptiSystem commercial tool. Different components, including PIN and APD photodiodes, were analyzed in the process of system analysis. DP-QPSK, 4-QAM, and 16-QAM modulation techniques were applied for system performance evaluation.

An architecture design of an (orthogonal frequency division multiplexing) OFDM-based coherent RoF system was proposed. The modeling was completed through OptiSystem, which is a commercial tool for optical components and system design. As reflected by the system analysis, the proposed OFDM-based coherent RoF system is able to undertake a high-speed backhaul network. System modeling was performed for 10 and > 100 Gb/s data rates. To measure the error vector magnitude (EVM) of visible light and the RoF hybrid system, an experiment

Hybrid Communication Systems for Future 6G and Beyond: Visible Light Communication & Radio over Fiber Technology, First Edition. Rao Kashif.

was conducted in the indoor scenario. The research results demonstrate that RoF can provide better communication, even in low lx scenarios.

The principles of uplink and channel switching in an ACSU were also discussed in this research, and the architecture of an ACSU was proposed. For more explicit comprehension, a simulation was performed in Xilinx ISE, with a 32-bit signal sent to the ACSU. The signal was switched to two different channels based on the control signal requirement. The control signal can be applicable to both a bandwidth-specific, application switching algorithm and a light-sensor, data-based switching algorithm. A light sensor data scenario was applied in the simulation. When the input data is 1, then the signal is shifted to VLC, whereas when the input data is 0, the signal is shifted to Wi-Fi. Since Wi-Fi was proposed as the uplink, an experiment regarding Wi-Fi signal over RoF and a power analysis were conducted under the conditions of a signal with and without digital predistortion (DPD). The results demonstrate that a Wi-Fi signal processed with DPD has better integrity in using signal power.

Indoor scenario modeling for the Hybrid VLC system was performed in OptiSystem, and the results indicate that RoF hybrid systems can increase the overall performance of a visible light system. Therefore, the proposed system can solve high-speed backhaul connections and uplink connectivity with channel switching in indoor VLC.

7.2 Future Work

In today's world, hybrid systems are in great demand because implementing a new technology can be easy if it can be integrated into current technologies. The integration of various technologies provides more benefits and lowers the cost of implementation. Researchers are already conducting future sixth-generation (6G) and seventh-generation (7G) research work, and according to some predictions, 6G will be an integration of satellite and optical networks, which will provide ultra-high speeds and a large flow of data communication. The proposed system can also handle the traffic for future hybrid networks, and with some system-level modification, this system can provide direct home broadband connectivity to broadband satellite networks. With regard to the ACSU, there should be a move from a system level toward a circuit level, and circuits and devices should be designed and made applicable in real indoor scenarios.

The area of visible light is still in its initial stages, and much research and system-level designs are required to make the technology applicable. This book-work is useful for people who are working in this area and want to contribute to indoor visible light and RoF communication. The book provides comprehensive knowledge of hybrid system designs for indoor VLC.

7.3 Applications of VLC in 6G and Above Communication

VLC has the potential to revolutionize communication systems, especially in the context of future generations like 6G and beyond. Here are some potential future applications of VLC in advanced communication systems.

7.3.1 High-Speed Data Transfer

In the context of 6G communication systems, VLC can contribute significantly to high-speed data transfer in several ways.

7.3.1.1 High Bandwidth

VLC utilizes the visible light spectrum, which offers a much broader bandwidth compared to radio frequencies currently used in wireless communication technologies like Wi-Fi or cellular networks. This wider bandwidth allows for higher data transfer rates, potentially enabling multi-gigabit-per-second transmission speeds.

7.3.1.2 Spectral Efficiency

VLC systems can achieve high spectral efficiency by employing techniques such as OFDM and multiple-input multiple-output (MIMO) communication. These techniques allow VLC systems to maximize data throughput by efficiently utilizing available spectrum resources.

7.3.1.3 Short-Range Communication

VLC is well suited for short-range communication scenarios, such as indoor environments or dense urban areas, where traditional wireless communication technologies may face interference or congestion issues. By leveraging VLC for short-range data transfer, 6G systems can provide high-speed connectivity in crowded environments without sacrificing performance.

7.3.1.4 Low Latency

VLC offers low-latency communication, making it suitable for applications that require real-time data transmission, such as virtual reality gaming, augmented reality applications, and industrial automation. In 6G networks, VLC can enable ultra-low latency communication, supporting emerging use cases that demand instantaneous response times.

7.3.1.5 Integration with Existing Infrastructure

VLC can leverage existing lighting infrastructure, such as light emitting diode (LED) luminaires, to establish communication links without the need for additional hardware deployment. This integration reduces deployment costs and accelerates the adoption of high-speed VLC-enabled communication systems in various environments.

7.3.1.6 Security and Privacy

VLC offers inherent security benefits, as visible light signals do not penetrate through walls or physical barriers, making eavesdropping difficult. In 6G networks, VLC can enhance security and privacy by providing secure communication channels for sensitive data transmission, such as financial transactions, medical records, and personal information.

7.3.1.7 Complementary to RF Technologies

VLC can complement traditional radio frequency (RF) communication technologies by offloading data traffic in congested or interference-prone environments. In 6G networks, VLC can serve as a complementary technology to RF-based systems, improving overall network capacity, reliability, and performance.

7.3.2 Indoor Localization and Navigation

VLC can significantly contribute to indoor localization and navigation systems in the following ways.

7.3.2.1 Precise Positioning

VLC systems utilize visible light signals emitted from LED luminaires to establish communication links with receiving devices, such as smartphones or wearable sensors. By measuring the signal strength and timing information from multiple VLC sources, indoor localization algorithms can accurately determine the position of the receiver within a confined space, enabling precise positioning with centimeter-level accuracy.

7.3.2.2 Multilayered Positioning

VLC systems can support multilayered positioning techniques by utilizing multiple floors or levels within a building. By deploying VLC transmitters on different floors, indoor localization systems can provide vertical positioning information, allowing users to navigate seamlessly between different levels of a building or indoor structure.

7.3.2.3 Low Latency

VLC offers low-latency communication, enabling real-time localization updates for dynamic indoor environments. By leveraging VLC for indoor positioning, navigation systems can provide instantaneous location updates to users, facilitating smooth and responsive navigation experiences, particularly in environments where quick decision-making is critical, such as emergency response scenarios or industrial automation applications.

7.3.2.4 High-Density Deployment

VLC systems can support high-density deployment scenarios, where multiple VLC transmitters are installed in close proximity to each other. By densely deploying VLC transmitters throughout indoor environments, localization and navigation systems can ensure continuous coverage and robust positioning capabilities, even in densely populated areas or complex indoor layouts.

7.3.2.5 Complementary to GPS

VLC complements existing localization technologies such as GPS, particularly in indoor environments where GPS signals may be weak or unavailable. By integrating VLC with GPS or other outdoor positioning systems, indoor localization and navigation systems can seamlessly transition between outdoor and indoor positioning modes, providing uninterrupted navigation services across different environments.

7.3.2.6 Integration with Smart Lighting

VLC-enabled smart lighting systems can serve dual purposes by providing both illumination and communication functionalities. By integrating VLC transmitters into LED luminaires, indoor lighting infrastructure can be utilized as communication nodes for indoor localization and navigation systems, enabling cost-effective deployment and seamless integration with existing lighting infrastructure.

7.3.2.7 Privacy and Security

VLC offers inherent privacy and security benefits, as visible light signals do not penetrate through walls or physical barriers. Indoor localization and navigation systems based on VLC can ensure user privacy by limiting location tracking to designated indoor spaces, thereby addressing privacy concerns associated with outdoor positioning technologies such as GPS.

7.3.3 Augmented Reality (AR) and Virtual Reality (VR)

VLC can enhance Augmented Reality (AR) and Virtual Reality (VR) experiences in several ways.

7.3.3.1 Low Latency Communication

VLC offers low-latency communication, allowing for real-time data transmission between AR/VR devices and content servers. This low latency is crucial for providing seamless and responsive AR/VR experiences, minimizing motion sickness, and enhancing user immersion.

7.3.3.2 High Bandwidth

VLC can provide high-speed data transfer, enabling the delivery of high-resolution AR/VR content with minimal compression artifacts. This high bandwidth facilitates the streaming of immersive multimedia content, such as 360° videos, virtual environments, and interactive simulations, enhancing the quality of AR/VR experiences.

7.3.3.3 Indoor Localization and Positioning

VLC can support indoor localization and positioning, allowing AR/VR devices to accurately track their position and orientation within indoor environments. By leveraging VLC signals emitted from LED lights or beacons, AR/VR systems can provide precise location-based information, enabling immersive augmented experiences that are tightly integrated with the user's physical surroundings.

7.3.3.4 Interactive Projection Mapping

VLC-enabled projection mapping systems can enhance AR experiences by overlaying virtual content onto physical surfaces in real-time. By synchronizing projected visuals with VLC signals emitted from light sources, AR systems can create dynamic and interactive visualizations that respond to user interactions and movements, blurring the line between virtual and physical environments.

7.3.3.5 Gesture Recognition

VLC can be utilized for gesture recognition and hand tracking in AR/VR applications. By analyzing changes in VLC signal patterns caused by hand movements or gestures, AR/VR systems can accurately detect and interpret user inputs, allowing for intuitive interaction with virtual objects and environments without the need for additional hardware sensors.

7.3.3.6 Privacy and Security

VLC offers inherent privacy and security benefits, as visible light signals are confined to specific physical areas and do not penetrate through walls or barriers. This characteristic makes VLC-based AR/VR systems suitable for privacy-sensitive applications, such as virtual meetings, training simulations, or confidential data visualization, where secure communication and data protection are paramount.

7.3.3.7 Multiuser Collaboration

VLC-based AR/VR systems can support multiuser collaboration in shared virtual environments. By establishing communication links between multiple AR/VR devices using VLC, users can collaborate in real-time, interact with shared virtual objects, and communicate with each other through voice or text chat, enhancing teamwork and productivity in collaborative AR/VR applications.

7.3.4 Smart Infrastructure and Internet of Things (IoT)

VLC has several applications in smart infrastructure and the Internet of Things (IoT), offering advantages such as high data rates, energy efficiency, and enhanced security. Below are some key applications.

7.3.4.1 Smart Lighting Systems

VLC can be integrated into LED lighting infrastructure to enable bidirectional communication between luminaires and IoT devices. Smart lighting systems using VLC can provide not only illumination but also data transmission capabilities, allowing for seamless integration with IoT sensors, actuators, and control systems. This integration enables advanced lighting control, energy management, and environmental monitoring in smart buildings, streets, and outdoor spaces.

7.3.4.2 Indoor Positioning and Navigation

VLC-based indoor positioning systems can support location-based services and navigation applications in smart buildings, airports, hospitals, and shopping malls. By deploying VLC transmitters throughout indoor environments, smart infrastructure can provide accurate positioning information to IoT devices, enabling indoor navigation, asset tracking, and personalized services for users.

7.3.4.3 Environmental Monitoring

VLC-enabled IoT sensors can collect environmental data, such as temperature, humidity, air quality, and light levels, for monitoring and control purposes in smart infrastructure. By leveraging VLC for data transmission, IoT sensors can communicate with centralized control systems efficiently, enabling real-time monitoring, analysis, and optimization of environmental conditions in buildings, urban areas, and industrial facilities.

7.3.4.4 Smart Retail and Hospitality

In retail and hospitality environments, VLC can enhance customer experiences and enable targeted advertising, wayfinding, and personalized services. By integrating VLC transmitters into lighting fixtures or display signage, smart infrastructure can deliver location-based information, promotional offers, and interactive content to customers' smartphones or wearable devices, enhancing engagement and loyalty.

7.3.4.5 Smart Transportation

VLC can support communication and data exchange between vehicles, roadside infrastructure, and transportation management systems in smart transportation networks. By deploying VLC transmitters on streetlights, traffic signals, and road signs, smart infrastructure can enable vehicle-to-infrastructure (V2I) communication, traffic monitoring, and congestion management, improving road safety and efficiency.

7.3.4.6 Industrial Automation

In industrial settings, VLC-based IoT systems can facilitate real-time monitoring and control of manufacturing processes, equipment, and inventory. By deploying VLC-enabled sensors and actuators in factories and warehouses, smart infrastructure can enable predictive maintenance, asset tracking, and optimization of production workflows, enhancing productivity and reducing downtime.

7.3.4.7 Energy Harvesting

VLC can be combined with energy harvesting technologies, such as solar panels or kinetic energy harvesters, to power IoT devices in smart infrastructure. By harvesting energy from ambient light or motion, VLC-enabled IoT sensors and actuators can operate autonomously without the need for batteries or external power sources, reducing maintenance costs and environmental impact.

7.3.5 Telecommunication/Wireless

VLC is a technology that uses visible light to transmit data wirelessly. It has garnered significant interest in recent years due to its potential applications in various fields, including future telecommunications. Here are some potential applications of VLC in future telecommunications.

7.3.5.1 Indoor Wireless Networking

VLC can be used for indoor wireless networking in environments where traditional RF communication may face interference or security concerns. By utilizing LED lights for data transmission, VLC can provide high-speed wireless connectivity in areas such as offices, hospitals, and schools.

7.3.5.2 Li-Fi

Li-Fi (light fidelity) is a subset of VLC that uses LEDs for data transmission. Li-Fi offers higher data rates compared to traditional Wi-Fi and can be deployed in scenarios where high-speed data transmission is crucial, such as in crowded urban areas or environments with a high density of connected devices.

7.3.5.3 Last-Mile Connectivity

VLC can serve as a solution for providing last-mile connectivity in areas where traditional wired or wireless infrastructure is unavailable or inadequate. By leveraging existing lighting infrastructure, VLC can extend connectivity to underserved or remote areas, bridging the digital divide.

7.3.5.4 Secure Communications

VLC offers enhanced security compared to RF communication technologies because light signals are confined within physical boundaries, making it difficult for eavesdroppers to intercept data. This makes VLC suitable for applications requiring secure communication, such as military and government networks, financial transactions, and healthcare systems.

7.3.5.5 Smart Cities

VLC can play a crucial role in the development of smart cities by enabling efficient data communication between various IoT devices, sensors, and infrastructure elements. Streetlights equipped with VLC technology can serve as data nodes, enabling smart street lighting, traffic management, environmental monitoring, and other smart city applications.

7.3.5.6 Augmented Reality (AR) and Location-Based Services

VLC can be used to deliver location-based services and augmented reality experiences in indoor environments where GPS signals may be unreliable or unavailable. By embedding location data in light signals, VLC can enable precise indoor positioning, navigation, and context-aware services.

7.3.5.7 Vehicle-to-Infrastructure (V2I) Communication

VLC can be employed for V2I communication, allowing vehicles to communicate with traffic lights, road signs, and other infrastructure elements using light signals. This can enhance road safety, traffic efficiency, and vehicle-to-vehicle communication in smart transportation systems.

8

The Role of AI and Machine Learning in 6G

8.1 Overview of AI and ML Concepts

Artificial intelligence (AI) involves creating machines that mimic human cognitive functions, enabling them to execute tasks that usually necessitate human intellect. These tasks encompass reasoning, learning, problem-solving, perception, and understanding language. Machine learning (ML), a branch of AI, concentrates on designing algorithms that enable computers to learn from data and make predictions or decisions accordingly.

8.1.1 Key AI and ML Concepts

(a) Supervised Learning
Supervised learning is a ML approach that involves training an algorithm on a dataset containing labeled examples. Each example in the dataset consists of an input paired with a corresponding output label. The goal is for the algorithm to learn a function that maps inputs to outputs, enabling it to accurately predict labels for new, unseen data.

I. Key Characteristics:
- Labeled Data: Requires a dataset with known outputs (labels).
- Training Process: Involves feeding the algorithm with the training data and adjusting the model parameters to minimize the error between predicted and actual labels.
- Applications in 6G:
  - o Traffic Classification: Recognizing various types of network traffic, such as video, voice, and data, to enhance resource allocation.

Hybrid Communication Systems for Future 6G and Beyond: Visible Light Communication & Radio over Fiber Technology, First Edition. Rao Kashif.

  - o Fault Detection: Using historical data to predict network issues or anomalies.
  - o User Behavior Prediction: Forecasting user demands and movement patterns to optimize the use of network resources.

II. Examples:
  - o Linear Regression: Estimating a continuous outcome based on input variables.
  - o Support Vector Machines (SVM): Categorizing data points into specified classes.
  - o Neural Networks: Understanding intricate relationships for applications such as image and speech recognition.

(b) Unsupervised Learning

Unsupervised learning deals with finding hidden patterns or intrinsic structures in data without labeled responses. The algorithm tries to learn the underlying structure of the data by clustering, dimensionality reduction, or association.

I. Key Characteristics:
  - No Labeled Data: Operates on datasets without explicit labels.
  - Pattern Discovery: Focuses on discovering hidden patterns, groupings, or features in the data.
  - Applications in 6G:
    - o Clustering: Grouping similar data points, such as user behaviors or network conditions, for targeted optimization.
    - o Anomaly Detection: Identifying unusual patterns that may indicate security threats or network faults.
    - o Dimensionality Reduction: Reducing the complexity of data for more efficient processing and visualization.

II. Examples:
  - K-Means Clustering: Partitioning data into k clusters based on similarity.
  - Principal Component Analysis (PCA): Reducing the number of variables while retaining most of the original variability.
  - Autoencoders: Neural networks used for unsupervised learning of efficient coding.

(c) Reinforcement Learning

Reinforcement learning (RL) is a type of ML where an agent learns to make decisions by performing actions in an environment to maximize cumulative reward. The agent receives feedback in the form of rewards or penalties and uses this feedback to improve its future actions.

I. Key Characteristics:

- Learning Through Interaction: The agent interacts with the environment and learns from the outcomes of its actions.
- Trial and Error: The learning process involves exploring different actions and learning from successes and failures.
- Applications in 6G:
  - o Dynamic Spectrum Allocation: Optimizing the use of available spectrum resources in real-time.
  - o Power Control: Adjusting transmission power to balance energy efficiency and signal quality.
  - o Network Slicing: Allocating resources dynamically to different network slices based on demand and priority.

II. Examples:
  - Q-Learning: A model-free RL algorithm that learns the value of actions based on expected future rewards.
  - Deep Q-Networks (DQN): Combining Q-learning with deep neural networks to handle high-dimensional state spaces.
  - Policy Gradient Methods: Learning policies that directly map states to actions to maximize expected rewards.

(d) Neural Networks

Neural networks are a set of algorithms inspired by the structure and function of the human brain. They are designed to recognize patterns and relationships in data through layers of interconnected nodes (neurons).

I. Key Characteristics:
  - Layered Structure: Consists of input, hidden, and output layers, where each layer contains multiple neurons.
  - Learning Through Weights: Neurons are connected by weights that are adjusted during training to minimize error.
  - Applications in 6G:
    - o Signal Processing: Enhancing signal quality and reducing noise.
    - o Image and Video Analysis: Processing and analyzing visual data for applications like security and augmented reality.
    - o Predictive Maintenance: Predicting network equipment failures to perform maintenance proactively.

II. Examples:
  - Feedforward Neural Networks (FNN): Basic neural networks where connections do not form cycles.

- Convolutional Neural Networks (CNN): Specialized for processing grid-like data, such as images.
- Recurrent Neural Networks (RNN): Designed for sequential data, such as time series or natural language.

8.2 Evolution of AI in Telecommunications

The integration of AI into telecommunications has profoundly transformed the industry, driving significant advancements in network management, optimization, and user experience. From its early beginnings with basic signal processing and rule-based systems to today's sophisticated deep learning models and real-time automation, AI's role has continuously expanded. As we approach the era of 6G, AI is poised to become even more integral, enabling unprecedented levels of network efficiency, reliability, and innovation. This chapter explores the evolutionary journey of AI in telecommunications, highlighting key milestones, technological advancements, and future trends shaping the next generation of wireless networks.

8.2.1 Early Adoption (1980s–1990s)

I. Key Applications:
   - Signal Processing: AI techniques like rule-based systems and simple neural networks were used for noise reduction, error correction, and signal enhancement.
   - Expert Systems: Employed in network management and diagnostics, these systems utilized predefined rules to troubleshoot and manage network issues.

II. Technological Landscape:
   - Limited Computational Power: Early AI systems were constrained by the available computing resources, leading to simpler, less efficient models.
   - Basic Algorithms: Use of basic ML algorithms such as linear regression, decision trees, and early neural networks.

8.2.2 Growth Phase (2000s)

I. Key Applications:
   - Network Management: AI was increasingly used for optimizing network performance, including traffic routing, load balancing, and resource allocation.
   - Customer Service: Introduction of AI-driven customer service tools like automated call centers and basic chatbots for handling common queries.

II. Technological Advancements:
   - Improved Algorithms: Development of more sophisticated ML algorithms, including SVM and early ensemble methods.
   - Data Availability: Increased availability of digital data from network operations and user interactions, providing a richer dataset for training AI models.

8.2.3 Modern Era (2010s)

I. Key Applications:
   - Predictive Maintenance: AI and ML were employed to predict network failures and perform proactive maintenance, reducing downtime and improving reliability.
   - Big Data Analytics: Leveraging large volumes of data to gain insights into network performance, user behavior, and service optimization.
   - Enhanced Customer Service: Advanced AI-driven chatbots and virtual assistants providing more sophisticated and accurate responses to user queries.

II. Technological Landscape:
   - Deep Learning: Emergence of deep learning techniques, such as CNNs and RNNs, enabling more complex and accurate models.
   - Cloud Computing: Use of cloud-based platforms for training and deploying AI models, providing the necessary computational power and scalability.
   - Internet of Things (IoT): Integration of AI with IoT devices for real-time monitoring and management of network resources.

8.2.4 Current Trends (2020s)

I. Key Applications:
   - 5G Networks: AI is integral to the operation and optimization of 5G networks, including dynamic spectrum management, network slicing, and enhanced mobile broadband (eMBB) services.
   - Network Automation: AI-driven automation for self-organizing networks (SON), reducing human intervention and operational costs.
   - Security and Fraud Detection: Advanced AI techniques for real-time threat detection, anomaly detection, and mitigation of security breaches.

II. Technological Advancements:
   - Edge AI: Deployment of AI capabilities at the edge of the network, enabling real-time data processing and decision-making closer to the data source.
   - Federated Learning: Collaborative ML across multiple decentralized devices or servers, preserving data privacy while improving model accuracy.

- Explainable AI (XAI): Development of AI models that provide transparent and understandable decision-making processes, crucial for regulatory compliance and user trust.

8.2.5 Future Directions (2030s and beyond)

I. Predicted Applications:
   - 6G Networks: AI will play a pivotal role in the development and operation of 6G networks, including terahertz communication, ultra-massive MIMO, and integrated sensing and communication.
   - Holographic and Tactile Internet: Enabling immersive experiences and real-time interaction through advanced AI-driven network capabilities.
   - Intelligent Automation: Full automation of network operations, driven by highly advanced AI systems capable of learning and adapting in real-time.

II. Technological Innovations:
   - Quantum Computing: Potential integration of quantum computing with AI to solve complex optimization problems at unprecedented speeds.
   - Advanced Neural Architectures: Development of more sophisticated neural network architectures, such as neural-symbolic integration and neuromorphic computing, mimicking human brain functions more closely.
   - AI Governance and Ethics: Establishment of robust frameworks for AI governance, addressing ethical considerations, data privacy, and regulatory compliance in telecommunications.

8.3 Why AI and ML are Critical for 6G

The integration of AI and ML into 6G networks is not just beneficial but essential to meet the ambitious goals and complex demands of the next-generation wireless communication systems. Here are the key reasons why AI and ML are critical for 6G.

I. Managing Complexity

Increased Network Complexity: 6G networks are expected to be significantly more complex than their predecessors, with the integration of diverse technologies such as terahertz communication, ultra-massive MIMO, and pervasive IoT devices. AI and ML can handle this complexity by automating network management tasks, analyzing vast amounts of data, and making real-time decisions that would be impossible for human operators to manage effectively.

Dynamic Environment: 6G networks will operate in highly dynamic environments with varying user demands, mobility patterns, and network conditions.

AI and ML algorithms can continuously learn and adapt to these changes, ensuring optimal network performance and reliability.

II. Enhancing Automation

Self-Organizing Networks (SON): AI-driven automation is key to the development of SON, where the network can self-configure, self-optimize, and self-heal with minimal human intervention. This reduces operational costs and improves network efficiency.

Predictive Maintenance: AI and ML can predict potential network failures or degradations before they occur, enabling proactive maintenance and reducing downtime. This predictive capability ensures higher reliability and better quality of service (QoS) for users.

III. Optimizing Resource Allocation

Efficient Spectrum Utilization: AI algorithms can dynamically allocate spectrum resources based on real-time analysis of network usage and interference patterns. This ensures efficient utilization of the available spectrum, which is crucial for meeting the high data rate and low latency requirements of 6G.

Energy Efficiency: AI and ML can optimize power consumption across the network by intelligently managing resources and predicting traffic loads. This is particularly important for sustainable network operation and reducing the carbon footprint of 6G networks.

IV. Enabling New Services and Applications

Ultra-Reliable Low-Latency Communications (URLLC): AI can optimize end-to-end communication paths to meet the stringent latency and reliability requirements of applications such as autonomous vehicles, remote surgery, and industrial automation.

Enhanced Mobile Broadband (eMBB): ML algorithms can enhance the user experience by predicting user behavior and preemptively allocating resources to ensure seamless connectivity and high data rates.

Massive Machine-Type Communications (mMTC): AI and ML can manage the massive number of IoT devices expected to be connected in 6G, ensuring efficient communication and coordination among devices.

V. Improving Security and Privacy

Real-Time Threat Detection: AI can enhance network security by detecting and responding to threats in real-time. ML models can identify unusual patterns and anomalies in network traffic, providing early warnings of potential attacks.

Privacy-Preserving Technologies: Techniques like federated learning enable collaborative model training without sharing raw data, preserving user privacy while still benefiting from collective learning across multiple devices or entities.

VI. Facilitating Advanced Network Features

Intelligent Beamforming: AI can optimize beamforming techniques to direct signals precisely where needed, improving signal quality and reducing interference in high-density environments.

Network Slicing: AI-driven network slicing allows the creation of multiple virtual networks on a shared physical infrastructure, each tailored to specific applications or user groups. This ensures that critical applications receive the necessary resources and QoS.

Context-Aware Services: AI and ML enable context-aware services that adapt to the user's environment and preferences, providing personalized and efficient communication experiences.

8.4 Applications of AI and ML in Wireless Networks

The deployment of AI and ML in wireless networks is revolutionizing the way these networks are managed, optimized, and utilized. Here are some key applications of AI and ML in wireless networks:

8.4.1 Network Management and Optimization

I. Self-Organizing Networks (SON):
- Self-Configuration: AI algorithms can automatically configure network parameters based on current conditions, reducing the need for manual setup.
- Self-Optimization: ML models can continuously monitor network performance and adjust settings to optimize key performance indicators (KPIs) such as throughput, latency, and coverage.
- Self-Healing: AI can detect and diagnose network issues in real-time, automatically initiating corrective actions to restore optimal performance.

II. Traffic Management:
- Traffic Prediction: AI can analyze historical data to predict traffic patterns, allowing for proactive management of network resources.
- Load Balancing: ML models can distribute network traffic evenly across available resources, preventing congestion and ensuring smooth operation.

III. Resource Allocation:
- Dynamic Spectrum Management: AI can optimize spectrum usage by dynamically allocating frequencies based on real-time demand and interference levels.

- Energy Efficiency: AI-driven techniques can optimize power consumption across the network, balancing performance with energy savings.

8.4.2 Enhanced User Experience

I. Quality of Service (QoS) Management:
  - Personalized Services: AI can tailor network services to individual user preferences and requirements, enhancing the user experience.
  - Predictive QoS: ML models can predict potential degradation in service quality and take preemptive measures to maintain high QoS.

II. Content Delivery:
  - Content Caching: AI can predict popular content and cache it closer to the end users, reducing latency and improving download speeds.
  - Adaptive Streaming: ML can optimize video streaming quality in real-time based on network conditions and device capabilities.

8.4.3 Security and Fraud Detection

I. Threat Detection:
  - Anomaly Detection: AI can identify unusual patterns in network traffic that may indicate security threats such as cyberattacks or unauthorized access.
  - Intrusion Prevention: ML models can predict potential security breaches and initiate preventive measures to protect network integrity.

II. Fraud Detection:
  - Pattern Recognition: AI can detect fraudulent activities by recognizing patterns in user behavior that deviate from the norm.
  - Real-Time Monitoring: ML algorithms can continuously monitor network transactions and flag suspicious activities for further investigation.

8.4.4 Predictive Maintenance and Fault Management

I. Predictive Analytics:
  - Failure Prediction: AI can analyze historical maintenance data and predict potential equipment failures, allowing for timely maintenance and reducing downtime.
  - Health Monitoring: ML models can monitor the health of network components in real-time, identifying signs of wear and tear before they lead to failures.

II. Automated Diagnostics:
  - Fault Detection: AI can diagnose network issues by analyzing performance data and identifying the root causes of problems.

- Automated Repair: ML algorithms can suggest or even implement corrective actions automatically, speeding up the repair process.

8.4.5 Advanced Communication Techniques

I. Beamforming and MIMO:
- Intelligent Beamforming: AI can optimize beamforming techniques to direct wireless signals precisely where needed, improving signal quality and reducing interference.
- Massive MIMO Optimization: ML can manage and optimize the complex antenna arrays in massive MIMO systems, enhancing network capacity and coverage.

II. Network Slicing:
- Dynamic Slicing: AI can create and manage virtual network slices tailored to specific applications or user groups, ensuring each slice meets its performance requirements.
- Resource Allocation: ML models can dynamically allocate resources to different slices based on real-time demand and service level agreements (SLAs).

8.4.6 Edge Computing and IoT

I. Edge Intelligence:
- Edge AI: Deploying AI capabilities at the edge of the network enables real-time data processing and decision-making, reducing latency and improving responsiveness.
- Federated Learning: AI models can be trained across multiple decentralized edge devices, preserving data privacy and improving model accuracy without centralizing data.

II. IoT Management:
- Device Management: AI can handle the vast number of IoT devices, ensuring efficient communication and coordination.
- Data Analytics: ML can analyze data generated by IoT devices to extract valuable insights and drive actionable outcomes.

8.5 6G and Visible Light Communication (VLC)

The convergence of 6G wireless communication technology with Visible Light Communication (VLC) holds tremendous potential for revolutionizing connectivity, enabling high-speed, secure, and energy-efficient data transmission. Here's how 6G and VLC intersect and the implications for future communication systems:

8.5.1 Ultrahigh-Speed Data Transmission

I. Combining Wireless and Optical Communication:
- 6G networks aim to achieve unprecedented data rates, leveraging advanced technologies such as terahertz communication and massive MIMO. VLC complements these efforts by providing additional bandwidth through optical frequencies, enabling ultra-high-speed data transmission.

II. Multi-Gigabit Data Rates:
- VLC can deliver multi-gigabit-per-second data rates, surpassing the capabilities of traditional wireless technologies. This enables applications such as ultra-high-definition video streaming, virtual reality, and augmented reality with minimal latency and high reliability.

8.5.2 Enhanced Indoor Localization and Positioning

I. Indoor Positioning Systems (IPS):
- VLC enables highly accurate indoor positioning systems by utilizing visible light sources such as light emitting diode (LED) bulbs and beacons. This technology can provide precise location information for indoor navigation, asset tracking, and location-based services.

II. Low Latency and High Accuracy:
- VLC-based IPS offers low-latency positioning with sub-meter accuracy, surpassing the limitations of GPS-based systems in indoor environments. This is particularly beneficial for applications requiring real-time localization, such as indoor robotics and augmented reality gaming.

8.5.3 Secure and Resilient Communication

I. Wireless Fidelity (Wi-Fi) Offloading:
- VLC can complement existing wireless technologies like Wi-Fi by offloading data traffic in densely populated areas, reducing congestion and improving network performance. This hybrid approach enhances connectivity while ensuring secure and reliable communication.

II. Secure Line-of-Sight Communication:
- VLC utilizes light as the communication medium, making it inherently secure against eavesdropping and interception. Line-of-sight transmission ensures that data is confined within the physical boundaries of the illuminated area, minimizing the risk of unauthorized access.

8.5.4 Energy-Efficient Networking

I. Green Communication Technologies:
- VLC operates using energy-efficient LEDs for data transmission, consuming significantly less power compared to traditional radio frequency

(RF) technologies. This contributes to reducing the carbon footprint of communication networks and promoting sustainable practices.

II. Integration with IoT Devices:
   - VLC is well-suited for communication with IoT devices due to its low-power requirements and compatibility with existing lighting infrastructure. This integration enables energy-efficient data exchange in smart homes, smart cities, and industrial IoT applications.

8.5.5 Overcoming RF Limitations and Interference

I. Spectrum Congestion Mitigation:
   - VLC operates in the unlicensed optical spectrum, alleviating the congestion often encountered in the crowded RF spectrum. This mitigates interference and enables reliable communication in environments where RF signals may be compromised.

II. Interference-Free Communication:
   - VLC offers interference-free communication in environments where RF signals may be susceptible to electromagnetic interference (EMI) or radio frequency interference (RFI), such as hospitals, aircraft cabins, and manufacturing facilities.

9

Future Research Directions for Visible Light Communication (VLC) in 6G Networks

As Visible Light Communication (VLC) continues to gain traction as a promising technology for wireless communication, especially in the context of 6G networks, several research directions emerge to address challenges and unlock its full potential. Here are some key areas for future research in VLC within the context of 6G.

9.1 VLC with Terahertz

9.1.1 Research Focus: Investigate Seamless Integration of VLC with Terahertz (THz) Communication Technologies

Objective: Exploit the complementary strengths of VLC in indoor environments and terahertz (THz) communication in outdoor scenarios to provide ubiquitous connectivity.

Overview: The next-generation wireless communication systems, particularly 6G, aim to achieve unprecedented levels of connectivity, speed, and reliability. VLC and THz communication are two promising technologies that can significantly contribute to these goals. VLC is well-suited for indoor environments due to its high bandwidth and secure line-of-sight (LOS) communication, while THz communication offers ultra-high data rates and is effective for outdoor and long-range communications. Integrating these two technologies can provide seamless and ubiquitous connectivity, leveraging the strengths of each.

9.1.1.1 Complementary Strengths

I. VLC Strengths:
- High bandwidth and data rates.
- Secure communication within confined areas.
- Low interference with radio frequency (RF) technologies.
- Utilization of existing lighting infrastructure.

Hybrid Communication Systems for Future 6G and Beyond: Visible Light Communication & Radio over Fiber Technology, First Edition. Rao Kashif.

II. THz Communication Strengths:
- Ultra-high data rates suitable for outdoor and long-distance communication.
- Large available spectrum in the THz band.
- Potential for high-resolution sensing and imaging applications.

9.1.1.2 Applications

- Indoor-Outdoor Connectivity: Provide seamless connectivity as users move from indoor to outdoor environments, ensuring uninterrupted high-speed communication.
- Smart Cities: Enable smart city applications where indoor VLC can be used in buildings, and THz communication can cover outdoor spaces, integrating sensors, vehicles, and public infrastructure.
- Industrial Automation: Use VLC for precise indoor positioning and control, while THz links provide high-speed communication between different sections of a factory or industrial site.

Challenges: Overcoming interface challenges, synchronization issues, and seamless handover mechanisms between VLC and THz links.

I. Interface Challenges:
- Heterogeneous Network Integration: Developing protocols and interfaces that allow VLC and THz communication systems to operate cohesively within the same network infrastructure.
- Hardware Compatibility: Ensuring that transceivers and network devices can support both VLC and THz communication standards without significant modifications.

II. Synchronization Issues:
- Time Synchronization: Achieving precise time synchronization between VLC and THz systems to ensure seamless data transfer and avoid communication delays.
- Frequency Coordination: Managing frequency synchronization to handle the transition between VLC and THz communication without loss of data integrity or performance.

III. Seamless Handover Mechanisms:
- Handover Algorithms: Developing intelligent handover algorithms that can predict the movement of users and preemptively switch between VLC and THz links to maintain continuous connectivity.
- Latency Management: Minimizing handover latency to ensure a smooth transition that is imperceptible to the user.
- Quality of Service (QoS): Maintaining consistent QoS during handover by dynamically adjusting resource allocation and managing interference.

9.1.1.3 Research Directions

I. Hybrid Network Architecture:
- Design a hybrid network architecture that integrates VLC and THz communication, ensuring seamless data flow between the two systems.
- Develop multi-mode transceivers capable of switching between VLC and THz modes based on environmental conditions and user requirements.

II. Cross-Layer Design:
- Implement cross-layer optimization techniques that consider physical, medium access control (MAC), and network layers to enhance the performance of VLC-THz integrated networks.
- Develop protocols that facilitate efficient coordination and resource sharing between VLC and THz communication layers.

III. Advanced Signal Processing:
- Utilize advanced signal processing algorithms to manage synchronization and handover processes.
- Explore machine learning techniques for predictive handover and adaptive resource management to enhance network performance.

IV. Simulation and Prototyping:
- Develop comprehensive simulation models to study the behavior and performance of integrated VLC-THz systems under various scenarios.
- Create prototypes and conduct real-world trials to validate theoretical models and optimize system parameters.

V. Security and Privacy:
- Address security challenges specific to the integration of VLC and THz communication, ensuring secure data transmission across both mediums.
- Develop privacy-preserving techniques to protect user data during handovers and across different communication links.

9.2 Enhanced Modulation and Coding Schemes

9.2.1 Research Focus: Develop Advanced Modulation and Coding Techniques Tailored for VLC in 6G Networks

Objective: Design modulation schemes optimized for the unique characteristics of visible light channels, such as flicker mitigation and color shift resilience.

Challenges: Addressing nonlinear distortion, interference from ambient light sources, and mitigating channel impairments in dynamic lighting conditions.

Overview: VLC leverages the visible spectrum to transmit data, which presents unique opportunities and challenges compared to traditional RF communication.

Advanced modulation and coding schemes are crucial for optimizing VLC performance, addressing the peculiarities of visible light channels, and meeting the high-speed and low-latency requirements of 6G networks.

9.2.1.1 Key Areas of Research

(a) Modulation Techniques: Developing modulation techniques that are robust against the inherent limitations and environmental influences on VLC systems is critical.

I. Orthogonal Frequency Division Multiplexing (OFDM):
- Advancements: Use of OFDM to efficiently utilize bandwidth and mitigate inter-symbol interference (ISI).
- Challenges: High Peak-to-Average Power Ratio (PAPR) in OFDM signals leading to nonlinear distortion in light emitting diode (LED) drivers.

II. Color Shift Keying (CSK):
- Advancements: Exploiting different colors (RGB) in LED-based VLC systems to encode data.
- Challenges: Managing color shift due to varying color perception and maintaining consistent illumination quality.

III. Pulse Amplitude Modulation (PAM):
- Advancements: Simple implementation and high energy efficiency.
- Challenges: Sensitivity to multipath interference and the need for precise timing synchronization.

IV. Discrete Multitone (DMT) Modulation:
- Advancements: High spectral efficiency and resilience to channel impairments.
- Challenges: Complexity in implementation and the need for sophisticated equalization techniques.

(b) Coding Techniques: Robust coding schemes are essential to correct errors and improve the reliability of VLC systems.

I. Forward Error Correction (FEC):
- Advancements: Implementation of powerful FEC schemes like LDPC (Low-Density Parity-Check) codes and Turbo codes.
- Challenges: Balancing the trade-off between coding gain and decoding complexity, especially for high-speed applications.

II. Visible Light Convolutional Coding (VLCC):
- Advancements: Tailored convolutional codes for VLC to handle burst errors and maintain data integrity.
- Challenges: Optimizing decoding algorithms to work efficiently in real-time scenarios.

III. Hybrid Automatic Repeat reQuest (HARQ):
- Advancements: Combining FEC with automatic repeat request (ARQ) for improved error correction and retransmission strategies.
- Challenges: Ensuring low latency in retransmission processes and managing increased overhead.

(c) Flicker Mitigation: VLC should not cause perceptible flicker, which can be annoying or even harmful to human eyes.

I. Constant Illumination:
- Advancements: Designing modulation schemes that maintain a constant average light intensity to avoid flicker.
- Challenges: Ensuring data encoding does not interfere with illumination quality, especially in dimming scenarios.

II. DC-Biased Optical OFDM (DCO-OFDM):
- Advancements: Using a direct current (DC) bias to ensure all transmitted signals are positive, thereby avoiding flicker.
- Challenges: Managing the power efficiency and ensuring linearity of the transmitted signal.

(d) Resilience to Ambient Light Interference: Ambient light, from sources such as sunlight and artificial lighting, can introduce significant noise and interference.

I. Adaptive Filtering:
- Advancements: Implementing adaptive filters to dynamically suppress ambient light noise.
- Challenges: Real-time adaptation to rapidly changing lighting conditions and maintaining signal integrity.

II. Optical Filtering:
- Advancements: Using optical filters to block out-of-band light frequencies that cause interference.
- Challenges: Designing filters that are effective yet do not significantly attenuate the desired signal.

(e) Mitigating Nonlinear Distortion: Nonlinear characteristics of LEDs and photodiodes can distort the transmitted and received signals.

I. Predistortion Techniques:
- Advancements: Applying predistortion to the input signal to counteract the nonlinear effects of LEDs.
- Challenges: Accurately modeling the nonlinearities and dynamically adjusting the predistortion parameters.

II. Digital Signal Processing (DSP):
- Advancements: Utilizing DSP techniques to correct nonlinear distortion post-reception.
- Challenges: Ensuring low-latency processing and maintaining high throughput.

(f) Dynamic Channel Impairments: VLC channels are highly variable due to changes in the environment, such as the movement of objects and changes in lighting conditions.

I. Channel Estimation and Equalization:
- Advancements: Developing robust channel estimation algorithms to adapt to varying channel conditions.
- Challenges: Ensuring accurate and fast channel estimation in real-time, and designing equalizers that can handle rapid changes.

II. Machine Learning for Channel Adaptation:
- Advancements: Employing machine learning techniques to predict and adapt to channel variations.
- Challenges: Training models that can generalize well across different environments and maintaining computational efficiency.

9.3 Hybrid VLC-RF Networks

9.3.1 Research Focus: Explore Hybrid Visible Light Communication (VLC) and Radio Frequency (RF) Network Architectures to Enhance Both Coverage and Reliability

Objective: Design efficient mechanisms that facilitate seamless switching between VLC and RF modes based on various factors such as channel conditions, mobility patterns, and specific application requirements.

9.3.1.1 Key Points

i. Channel Conditions:
- VLC performance is influenced by light source visibility, ambient light interference, and LOS conditions.
- RF performance can be affected by physical obstacles, interference, and signal fading.
- Efficient mechanisms must assess and react to these varying conditions to maintain optimal connectivity.

II. Mobility Patterns:
- Devices and users move frequently, impacting the stability of both VLC (which requires LOS) and RF connections.

- Mechanisms must predict and adapt to mobility, ensuring continuity and QoS.

III. Application Requirements:
- Different applications have unique demands in terms of bandwidth, latency, and reliability.
- Mechanisms must cater to these diverse requirements, providing tailored connectivity solutions.

9.3.1.2 Challenges

i. Developing Intelligent Handover Algorithms:
- Seamless Transition: Algorithms must enable smooth transitions between VLC and RF without noticeable interruptions or degradation in service.
- Context-Awareness: Algorithms must understand and react to the current context, including user behavior, network conditions, and application needs.
- Proactivity: Anticipating the need for handovers before the current channel degrades to an unacceptable level.

ii. Optimizing Resource Allocation:
- Dynamic Allocation: Resources must be dynamically allocated between VLC and RF to maximize overall network performance.
- Load Balancing: Efficiently distributing traffic across VLC and RF channels to avoid congestion and ensure balanced usage of available resources.
- Energy Efficiency: Minimizing power consumption, particularly for battery-operated devices, by smartly switching between VLC and RF.

iii. Managing Interference:
- Coexistence: Ensuring that VLC and RF technologies can coexist without causing detrimental interference to each other.
- Interference Mitigation: Developing techniques to mitigate both intra-system (within VLC or RF) and inter-system (between VLC and RF) interference.
- Adaptive Techniques: Implementing adaptive modulation, coding schemes, and frequency/channel selection to manage interference dynamically.

9.3.1.3 Potential Solutions and Approaches

I. Machine Learning and AI:
- Utilizing machine learning algorithms to predict channel conditions and user mobility, thereby optimizing handovers and resource allocation.
- AI-driven decision-making for real-time adaptation to network changes.

II. Advanced Handover Protocols:
- Developing new handover protocols specifically designed for hybrid VLC-RF environments.

- Protocols that incorporate predictive analytics to preemptively manage handovers.

III. Cross-Layer Optimization:
- Integrating mechanisms across multiple layers of the network stack (physical, MAC, network) to ensure holistic optimization.
- Ensuring that changes in one layer positively impact overall network performance.

9.3.1.4 Collaborative Communication Strategies

- Leveraging collaboration between VLC and RF modules to enhance network robustness and reliability.
- Employing cooperative communication techniques, such as relaying and network coding.

9.4 Massive MIMO and Beamforming Techniques

9.4.1 Research Focus: Investigate the Integration of Massive Multiple-Input Multiple-Output (MIMO) and Beamforming Techniques Within Visible Light Communication (VLC)-Enabled 6G Networks

Objective: Harness the spatial diversity and beamforming capabilities of massive multiple-input multiple-output (MIMO) to significantly enhance coverage, capacity, and spectral efficiency in VLC systems.

9.4.1.1 Key Points

I. Spatial Diversity:
- Utilizing a large number of LEDs and photodetectors to exploit spatial dimensions.
- Enhancing link reliability by providing multiple propagation paths between the transmitter and receiver.

II. Beamforming Capabilities:
- Steering light beams toward specific users to focus energy where it is needed most.
- Improving signal strength and quality by dynamically adjusting the direction and shape of the light beams.

III. Coverage Enhancement:
- Extending the effective coverage area of VLC systems by leveraging the directional properties of beamforming.
- Overcoming LOS limitations inherent to VLC.

IV. Capacity Improvement:
- Increasing data rates by using multiple data streams simultaneously via Massive MIMO.
- Exploiting spatial multiplexing to enhance system throughput.

V. Spectral Efficiency:
- Efficiently utilizing the available spectrum by minimizing interference and optimizing the allocation of resources.
- Enabling high-density deployment scenarios typical in 6G networks.

9.4.1.2 Challenges

I. Scalability of Massive MIMO in VLC Deployments:
- Hardware Complexity: Managing the complexity and cost associated with deploying large arrays of LEDs and photodetectors.
- Heat Dissipation: Addressing the thermal management issues due to the dense packing of light sources.
- Synchronization: Ensuring precise synchronization among a large number of transmitters and receivers to maintain coherent signal processing.

II. Optimizing Beamforming Algorithms for Indoor Environments:
- Dynamic Environments: Adapting to changing indoor environments with moving objects and varying reflectivity.
- Non-LOS Conditions: Developing robust beamforming techniques that can handle non-LOS scenarios caused by obstacles.
- User Mobility: Tracking and adjusting beam directions in real-time to accommodate user movement.

III. Mitigating Inter-Cell Interference:
- Dense Deployments: Managing interference in environments with a high density of VLC cells.
- Co-channel Interference: Developing strategies to minimize interference between adjacent cells operating on the same frequencies.
- Interference Coordination: Implementing coordination mechanisms among cells to dynamically adjust power levels and beam patterns.

9.4.1.3 Potential Solutions and Approaches

I. Advanced Beamforming Algorithms:
- Hybrid Beamforming: Combining analog and digital beamforming techniques to achieve a balance between performance and complexity.
- Machine Learning: Utilizing machine learning to predict and adapt to dynamic indoor environments, optimizing beam directions and power allocation.

- Adaptive Beamforming: Implementing algorithms that can dynamically adjust beam patterns in response to real-time changes in the environment and user locations.

II. Massive MIMO Architectures:

- Modular Designs: Developing scalable and modular Massive MIMO architectures to simplify deployment and maintenance.
- Distributed MIMO: Exploring distributed MIMO setups where multiple smaller MIMO arrays are spread across the coverage area to improve scalability and reduce complexity.

III. Interference Mitigation Techniques:

- Coordinated Multipoint (CoMP): Leveraging coordination among multiple VLC cells to mitigate inter-cell interference and enhance overall system performance.
- Dynamic Spectrum Management: Implementing dynamic spectrum allocation and interference management strategies to adapt to varying traffic demands and environmental conditions.
- Power Control: Using intelligent power control mechanisms to minimize interference while maintaining adequate signal quality.

9.5 Network Slicing and Service Differentiation

9.5.1 Research Focus: Explore Network Slicing and Service Differentiation Mechanisms Tailored for Visible Light Communication (VLC) Networks Within the Context of 6G

Objective: Design dynamic network slicing architectures that efficiently allocate resources based on diverse application requirements, such as indoor localization, high-speed data transmission, and low-latency communication.

9.5.1.1 Key Points

I. Network Slicing:

- Creating multiple virtual networks (slices) on a single physical infrastructure.
- Each slice is tailored to meet specific application requirements and performance metrics.
- Enhancing flexibility and efficiency in resource utilization by isolating different types of traffic.

II. Service Differentiation:

- Categorizing network services based on their unique needs, such as bandwidth, latency, and reliability.
- Ensuring that each service category receives appropriate resources and QoS.

9.5.1.2 Challenges

I. Developing Scalable Slicing Frameworks:
- Resource Management: Allocating physical and virtual resources dynamically and efficiently across different slices.
- Orchestration: Coordinating the creation, management, and tear-down of network slices in real-time.
- Scalability: Ensuring that the slicing framework can scale to support a large number of diverse applications and devices.

II. Defining Service-Level Agreements (SLAs):
- Customization: Defining SLAs that cater to the specific needs of various applications, including stringent requirements for latency, throughput, and reliability.
- Monitoring and Enforcement: Continuously monitoring network performance to ensure SLAs are met, and taking corrective actions when they are not.
- Flexibility: Allowing dynamic adjustment of SLAs based on changing network conditions and application requirements.

III. Ensuring Isolation and Security Between Different Slices:
- Isolation: Preventing interference and ensuring that the performance of one slice does not impact others.
- Security: Protecting data and resources within each slice from unauthorized access and attacks.
- Resource Isolation: Implementing mechanisms to guarantee that each slice receives its allocated resources without contention.

9.5.1.3 Potential Solutions and Approaches

I. Advanced Orchestration Platforms:
- Software-Defined Networking (SDN): Using SDN controllers to manage and orchestrate network slices dynamically.
- Network Function Virtualization (NFV): Deploying virtualized network functions that can be allocated to different slices as needed.
- AI and Machine Learning: Leveraging AI and machine learning for predictive resource allocation and optimization.

II. Dynamic Resource Allocation Mechanisms:
- Adaptive Resource Management: Developing algorithms that can adaptively allocate resources based on real-time network conditions and application demands.
- Multidimensional Resource Allocation: Considering multiple dimensions such as bandwidth, power, and computational resources in the allocation process.

III. Enhanced Security and Isolation Techniques:
- Cryptographic Isolation: Using encryption to ensure that data within a slice is accessible only to authorized users.
- Access Control: Implementing strict access control mechanisms to regulate who can create, modify, or access different slices.
- Interference Management: Developing techniques to minimize interference between slices, such as through advanced scheduling and resource partitioning.

9.5.1.4 Application Scenarios

I. Indoor Localization:
- Low-Latency Requirements: Ensuring that the network slice dedicated to indoor localization provides ultra-low latency for real-time tracking.
- High Accuracy: Allocating resources to maintain high accuracy in positioning and tracking.

II. High-Speed Data Transmission:
- High Throughput: Ensuring that slices intended for high-speed data transmission are allocated sufficient bandwidth and processing power.
- Reliability: Guaranteeing reliable data delivery, even in dense environments with many users.

III. Low-Latency Communication:
- Real-Time Applications: Supporting applications such as augmented reality (AR) and virtual reality (VR) that require extremely low latency.
- Quality of Service: Maintaining consistent QoS to prevent latency spikes and ensure smooth user experiences.

9.6 Energy-Efficient VLC Systems

9.6.1 Research Focus: Develop Energy-Efficient Visible Light Communication (VLC) Systems Tailored for Sustainable 6G Networks

Objective: Design low-power VLC transmitters, receivers, and Internet of Things (IoT) devices to minimize energy consumption and extend battery life.

9.6.1.1 Key Points

I. Low-Power VLC Transmitters:
- Designing LED drivers and circuits that operate efficiently at low power levels.

- Utilizing energy-efficient modulation schemes to reduce power consumption.

II. Energy-Efficient Receivers:
- Developing photodetectors and associated circuitry that consume minimal power while maintaining high sensitivity.
- Implementing low-power signal processing algorithms.

III. Energy-Saving IoT Devices:
- Creating IoT devices with integrated VLC capabilities that prioritize energy efficiency.
- Optimizing sleep/wake cycles and power management techniques.

9.6.1.2 Challenges

I. Optimizing Power Allocation:
- Dynamic Power Control: Implementing adaptive power control mechanisms that adjust the power levels of VLC transmitters based on real-time conditions.
- Efficient Power Distribution: Ensuring optimal power distribution among multiple LEDs in a VLC system to minimize overall energy consumption.

II. Modulation Schemes for Energy Efficiency:
- Low-Power Modulation Techniques: Developing and utilizing modulation schemes that require less power while maintaining reliable communication.
- Adaptive Modulation: Employing adaptive modulation techniques that switch between different schemes based on the channel conditions to balance energy consumption and performance.

III. Signal Processing Algorithms:
- Energy-Efficient Algorithms: Designing signal processing algorithms that perform complex tasks with minimal energy expenditure.
- Algorithm Optimization: Optimizing existing algorithms for lower computational complexity and power consumption.

9.6.1.3 Potential Solutions and Approaches

I. Advanced LED Driver Design:
- Efficient Circuits: Creating driver circuits that optimize the current flow to LEDs, reducing energy wastage.
- Pulse Width Modulation (PWM): Utilizing PWM techniques to adjust the brightness and power usage of LEDs dynamically.

II. Energy-Aware Modulation Techniques:
- Optical OFDM: Implementing Optical OFDM with energy-efficient coding schemes.

- Dimming-Compatible Modulation: Designing modulation schemes that are compatible with dimming functionalities to save energy.

III. Low-Power Signal Processing:
- Analog Signal Processing: Exploring analog signal processing techniques that can be more energy-efficient compared to digital processing.
- Sparse Signal Processing: Utilizing sparse signal processing algorithms to reduce the number of active components and operations, thus saving energy.

IV. Efficient Network Protocols:
- Sleep Mode Protocols: Developing protocols that allow devices to enter low-power sleep modes during periods of inactivity.
- Energy-Aware MAC Protocols: Implementing MAC protocols that optimize energy usage by minimizing collisions and idle listening.

V. Optimized IoT Device Design:
- Low-Power Hardware: Designing hardware components specifically for low energy consumption, including energy-efficient processors and sensors.
- Energy Harvesting: Integrating energy harvesting techniques to supplement battery power, such as solar cells that can convert ambient light into electrical energy.

9.6.1.4 Application Scenarios

I. Smart Homes and Buildings:
- Lighting Control: Using VLC for intelligent lighting systems that adjust brightness and power consumption based on occupancy and ambient light levels.
- Energy Management: Integrating VLC with energy management systems to optimize power usage across various devices.

II. Healthcare and Wearables:
- Medical IoT Devices: Designing wearable medical devices that use VLC for data transmission while minimizing energy usage to extend battery life.
- Patient Monitoring: Implementing VLC in patient monitoring systems that require low power consumption to ensure long-term operation without frequent battery replacements.

III. Industrial Automation:
- Energy-Efficient Sensors: Developing VLC-enabled sensors for industrial automation that consume minimal power, enhancing the sustainability of smart factories.
- Machine-to-Machine Communication: Utilizing energy-efficient VLC for reliable communication between machines, reducing overall power consumption in industrial settings.

9.7 Security and Privacy Enhancements

9.7.1 Research Focus: Investigate Advanced Security and Privacy Mechanisms Specifically Designed for Visible Light Communication (VLC) in 6G Networks

Objective: Develop robust authentication, encryption, and privacy-preserving techniques to protect VLC transmissions from threats such as eavesdropping, spoofing, and tampering.

9.7.1.1 Key Points

I. Robust Authentication:
- Ensuring that only authorized devices can access and communicate over the VLC network.
- Preventing unauthorized entities from intercepting or injecting data into the communication channel.

II. Encryption:
- Implementing strong encryption algorithms to protect data transmitted via VLC.
- Ensuring data confidentiality and integrity during transmission.

III. Privacy-Preserving Techniques:
- Protecting user privacy by anonymizing communication metadata.
- Ensuring that personal and sensitive information is not exposed through VLC communications.

9.7.1.2 Challenges

I. Line-of-Sight (LOS) Requirements:
- Vulnerability to Eavesdropping: VLC's LOS nature makes it easier for an attacker within the line of sight to intercept signals.
- Restricted Coverage: Ensuring secure communication in environments where LOS can be easily obstructed.

II. Physical-Layer Attacks:
- Jamming: Deliberate interference with VLC signals to disrupt communication.
- Spoofing: Attacks where a malicious entity impersonates a legitimate device to gain unauthorized access.
- Tampering: Physical alteration of VLC devices or light sources to compromise security.

III. Visible Light Signal Risks:
- Exposure: The visible nature of VLC signals can potentially expose information to unintended observers.

- Environmental Interference: Managing security in environments with varying light conditions and potential signal reflections.

9.7.1.3 Potential Solutions and Approaches

I. Advanced Authentication Mechanisms:
   - Two-Factor Authentication (2FA): Combining VLC with another authentication method (e.g. RF-based) to enhance security.
   - Physical Unclonable Functions (PUFs): Using unique physical characteristics of VLC hardware for secure device authentication.
   - Cryptographic Handshakes: Implementing strong cryptographic protocols to verify the identity of communicating devices.

II. Strong Encryption Algorithms:
   - Lightweight Cryptography: Developing encryption algorithms optimized for low-power VLC devices.
   - Quantum Cryptography: Exploring quantum key distribution (QKD) for ultra-secure VLC communications.
   - End-to-End Encryption: Ensuring that data is encrypted at the source and only decrypted at the destination.

III. Privacy-Preserving Techniques:
   - Anonymous Communication Protocols: Implementing protocols that hide user identities and communication patterns.
   - Data Minimization: Reducing the amount of personal data transmitted over VLC networks to limit exposure.
   - Secure Multiparty Computation (SMC): Allowing multiple parties to jointly compute a function over their inputs while keeping those inputs private.

IV. Physical-Layer Security Measures:
   - Directional Transmission: Using beamforming to direct VLC signals precisely toward intended receivers, minimizing leakage.
   - Optical Fencing: Implementing physical barriers that block VLC signals from reaching unintended areas.
   - Signal Obfuscation: Introducing controlled noise or dummy data to confuse potential eavesdroppers without degrading legitimate communication.

V. Environment-Adaptive Security:
   - Dynamic Key Management: Regularly updating encryption keys based on environmental changes to maintain security.
   - Context-Aware Security Policies: Adapting security measures based on real-time assessment of the environment and potential threats.
   - Reflection and Interference Mitigation: Techniques to detect and mitigate risks associated with signal reflections and ambient light interference.

9.7.1.4 Application Scenarios

I. Smart Homes and Offices:
   - Secure Lighting Systems: Ensuring that smart lighting solutions using VLC are protected against unauthorized access and control.
   - Confidential Communication: Enabling secure communication channels within homes and offices for sensitive data transmission.

II. Healthcare:
   - Patient Data Privacy: Protecting patient data transmitted via VLC in healthcare facilities to comply with privacy regulations.
   - Secure Medical IoT Devices: Ensuring that medical devices using VLC for communication are secure from tampering and unauthorized access.

III. Industrial Automation:
   - Securing Machine-to-Machine (M2M) Communication: Protecting data exchanged between machines in industrial environments to prevent espionage and sabotage.
   - Environmental Adaptation: Implementing security measures that can adapt to the dynamic and often harsh industrial settings.

Index

Hybrid Communication Systems for Future 6G and Beyond: Visible Light Communication & Radio over Fiber Technology, First Edition. Rao Kashif.

Printed and bound by CPI Group (UK) Ltd, Croydon, CR0 4YY

26/03/2025

14648614-0005